Springer Tracts in Modern Physics
Volume 236

Springer Tracts in Modern Physics

Springer Tracts in Modern Physics provides comprehensive and critical reviews of topics of current interest in physics. The following fields are emphasized: elementary particle physics, solid-state physics, complex systems, and fundamental astrophysics.
Suitable reviews of other fields can also be accepted. The editors encourage prospective authors to correspond with them in advance of submitting an article. For reviews of topics belonging to the above mentioned fields, they should address the responsible editor, otherwise the managing editor.
See also springer.com

Gary John Barker

b-Quark Physics with the LEP Collider

The Development of Experimental Techniques for b-Quark Studies from Z^0-Decay

Springer

Dr. Gary John Barker
University of Warwick
Dept. Physics
Gibbet Hill Road
Coventry
United Kingdom CV4 7AL
g.j.barker@warwick.ac.uk

G.J. Barker, *b-Quark Physics with the LEP Collider: The Development of Experimental Techniques for b-Quark Studies from Z^0-Decay*, STMP 236 (Springer, Berlin Heidelberg 2010), DOI 10.1007/978-3-642-05279-8

ISSN 0081-3869 e-ISSN 1615-0430
ISBN 978-3-642-05278-1 e-ISBN 978-3-642-05279-8
DOI 10.1007/978-3-642-05279-8
Springer Heidelberg Dordrecht London New York

Library of Congress Control Number: 2010922801

Cover design: Integra Software Services Pvt. Ltd., Pondicherry

Printed on acid-free paper

Springer is part of Springer Science+Business Media (www.springer.com)

For Sabine

Preface

The b-physics results from the LEP data sets exceeded all expectations in both the range of topics covered and the precision of measurements that were eventually possible. The success was due to several factors such as the rapid development of silicon-strip vertex detectors, advances in the theoretical description of b-hadrons and not least because of the inventiveness, skill and determination of the experimenters performing the data analysis. In order to utilise fully the precision of the new vertex detectors, careful commissioning and event reconstruction was needed. This together with new methods and techniques of data analysis, particularly in the area of inclusive b-hadron reconstruction, were the key to a whole host of measurements that spanned the entire field of b-physics.

The aim of this book is not to give another review of b-physics results. Instead the focus is on reviewing the main experimental methods that evolved (hardware and software) and the lessons learnt from somebody who was involved first-hand in their development and use. The hope is that this work will both stand as a record of what was achieved with the LEP data, and also as a reference source of b-physics analysis ideas for experimenters embarking on projects in the future e.g. at the LHC or a linear collider.

Results from all four of the LEP experiments are reported but some bias towards DELPHI is inevitable, especially to illustrate general points, since I am a collaboration member and have easier access to material. I wish to thank my friends and colleagues of the DELPHI collaboration for making the project such a success.

In addition I wish to give special thanks to: Prof. Dr. Michael Feindt who was the instigator of inclusive analysis methods for b-physics with the LEP data; Prof. Dr Thomas Müller for giving me the opportunity and encouragement to write this article; Dr. Richard Hawkings (OPAL) and Dr. Christian Weiser (DELPHI) for allowing me access to some unpublished results and for useful discussions. Finally I would like to thank the people at Springer, especially Ute Heuser for her professional handling of the project and patience with my regular failure to meet deadlines.

University of Warwick, UK
September 2009

Gary J. Barker

Contents

Chapter 1
b-physics at LEP

1.1 Introduction

The LEP project [1] was conceived following the release of a report in 1976 by Burton Richter [2] calling for the development of an e^+e^--machine to study the nature of the weak interaction. A physics program began to be defined in the late 1970s where the study of b-quarks was mentioned in the context of electroweak measurements (R_b, A_{FB}^{b}) but 'b-physics' was not discussed as a research line in its own right. This was reasonable since the b-quark lifetime, although at the time unknown, was expected to be so short[1] that the separation of events containing b-quarks from the rest, would be difficult.

Two developments changed this situation: The first was in 1982 and 1983 when the MAC and MARKII experiments at the PEP collider (30 GeV e^+e^- collisions) started to see that the lifetime of b-hadrons was about a factor ten larger than expected [3, 4]. The second, was the development at about this time of silicon-strip detectors and their successful operation in fixed-target experiments. This provided, in principle, the means by which the decay vertices of 'long-lived' b-particles could be reconstructed and two of the four LEP experiments, ALEPH and DELPHI, included designs for silicon-strip vertex detectors in their original letters of intent in 1982. We review the technology and performance of the LEP vertex detectors in Chap. 2.

In 1989, just before LEP was switched on for the first time, a CERN yellow report [5] outlined what b-physics goals might reasonably be achieved from a data sample of 3×10^6 b-hadrons. The main conclusions were:

- An accuracy of ± 0.001 on $\sin\theta_W$ by measuring the forward-backward asymmetry of b-jets.
- Testing the validity of b-quark fragmentation models.

---

[1] Based on the assumption that the coupling between the second and third generation of fermions was of roughly the same size as that measured between the first and second generation.

G.J. Barker, *b-Quark Physics with the LEP Collider*, STMP 236, 1–21,
DOI 10.1007/978-3-642-05279-8_1,

- Perhaps measuring separately the lifetimes of the different weakly decaying b-hadron states if the differences between them were as high as 30%.
- $B\bar{B}$ mixing measurements:

  (a) the time integrated variables $\bar{\chi}$ with 15% precision and χ_s with 25% statistical precision from 10^6 and 10^7 Z^0-decays respectively,
  (b) expecting to confirm the time evolution of B^0_d mixing but with B^0_s oscillation being too fast for detection

- The value of $|V_{ub}/|V_{cb}|$ constrained below 0.21.

As it turned out, these goals were conservative and the final LEP b-physics results surpassed all expectations. A major contributing factor was that many estimates were based on applying the analysis methods and techniques that had been successful at the PEP and PETRA experiments. Here, the presence of high P_t leptons and D-mesons had been important signatures of b-quarks. At LEP, the relatively large decay length of b-hadrons and the use of precision vertex detectors meant that additional, more powerful, *lifetime* signatures of b-quarks could be used. This 'tagging' of b-quark events at LEP is described in Chap. 4. Advanced particle identification techniques coupled with modern combined-variable statistical methods, enabled the development of *inclusive* b-hadron reconstruction which compensated for the otherwise low statistics expected in exclusive decay channels of b-hadrons at the Z^0. In Chap. 7 we highlight the progress made in this area and Chap. 6 covers the use of double-tag methods which were crucial to the final precision attained on many results due to a reduced dependence on simulations.

Running in parallel to these experimental developments was a near revolution in the predictive power of heavy quark theory. Many areas of the LEP b-physics program benefitted from application of the heavy quark symmetry and some long-standing theoretical limiting uncertainties were removed. These ideas are briefly described in Sect. 1.3.4.

Progress during this period was also aided enormously by the compilation and averaging of results by the LEP working group structure. The Electroweak Working Group (EWWG) and the Heavy Flavour Steering Group (later the Heavy Flavour Averaging Group, HFAG) played an essential role in combining electroweak and non-electroweak results from the different experiments and in defining common frameworks for reporting results to aid and improve this process. Some further details are given in Sect. 1.4.

Finally, we emphasise that this book is primarily an account of the experimental developments which enabled the LEP b-physics program to be such an enormous success. Although many of the results that were the highlights of the program are featured to illustrate experimental methods, it is not a complete reference source of LEP b-physics results. The myriad of results that form the LEP legacy have been amply documented elsewhere and we refer the reader to a sample of some of the compilations that are available e.g. [6–8] and [9].

1.2 The LEP Machine and Experiments

The LEP accelerator was operated between 1989 and 1995 at a collision energy of $\sim$ 90 GeV (LEP I) in order to produce Z^0 bosons. Collisions occurred at four points around the ring and each point was instrumented with a large, multi-purpose particle detector. By the end of LEP I, the collider had supplied about four million Z^0-bosons to each of the experiments. This first phase of LEP operation, was followed in 1996 by a period of higher energy running (LEP II) where the machine energy was tuned to study $W^{\pm}$-pair and Z^0-pair production. By the time the plug was pulled on the LEP project in 2000, the collision energy had reached 208 GeV. Essentially all of the b-physics program was performed using data from the LEP I data.

The four LEP detectors: ALEPH [10, 11], DELPHI [12, 13], L3 [14] and OPAL [15] were a mixture of tried-and-tested technology and bold new state-of-the-art detection methods. The detectors were first and foremost multipurpose devices instrumented to record all products of $e^+e^- \rightarrow Z^0$ interactions. This largely constrained all designs to follow the 'onion layer' format of: tracking detectors at the centre (closest to the interaction point) to measure particle trajectories, surrounded by electromagnetic and hadronic calorimeters to measure particle energy and chambers around the outside to detect the penetrating muons. This format is illustrated by the layout of the OPAL detector shown in Fig. 1.1.

The tracking detectors were placed within a solenoidal magnetic field aligned with the beam-line and consisted of silicon-strip detectors placed immediately surrounding the beampipe to measure the decay vertices of weakly decaying c-and b-hadrons. Outside of the silicon came large-scale gaseous ionisation detectors (wire drift chambers or time projection chambers) whose primary function was to reconstruct the curved trajectories (or *tracks*) of charged particles in the magnetic field in order to measure their momentum.

Outside of the solenoid the calorimeters would be typically split into a cylindrical part and two end-cap devices in order to cover the full solid angle. The electromagnetic calorimetry recorded the electromagnetic showers instigated by electrons and photons and surrounding this, typically instrumenting the iron return yolk for the magnetic field, was the hadronic calorimeter registering hadronic showers.

The process $e^+e^- \rightarrow Z^0 \rightarrow b\bar{b}$ is seen in the detectors as a system of *jets* associated with the hadronisation (see Sect. 1.3.5) of the $b\bar{b}$ quark pair and of any associated high p_t gluon that may be present. Figure 1.2 shows a typical 3-jet event reconstructed by the OPAL detector in the $R - \phi$ projection.[2] Tracks can be seen streaming out from the central e^+e^- interaction point and blocks register the location of energy deposits in the time-of-flight, electromagnetic and hadronic calorimeters respectively, whose size is proportional to the energy deposit. The arrow represents a muon candidate registered in the muon chambers.

---

[2] The natural choice of coordinate system is cylindrical polar coordinates (R, ϕ, z) with the z-axis aligned along the e^+e^- beam-line so that charged particles bend under the magnetic field in the $(R - \phi)$ plane i.e. the plane transverse to the z-axis.

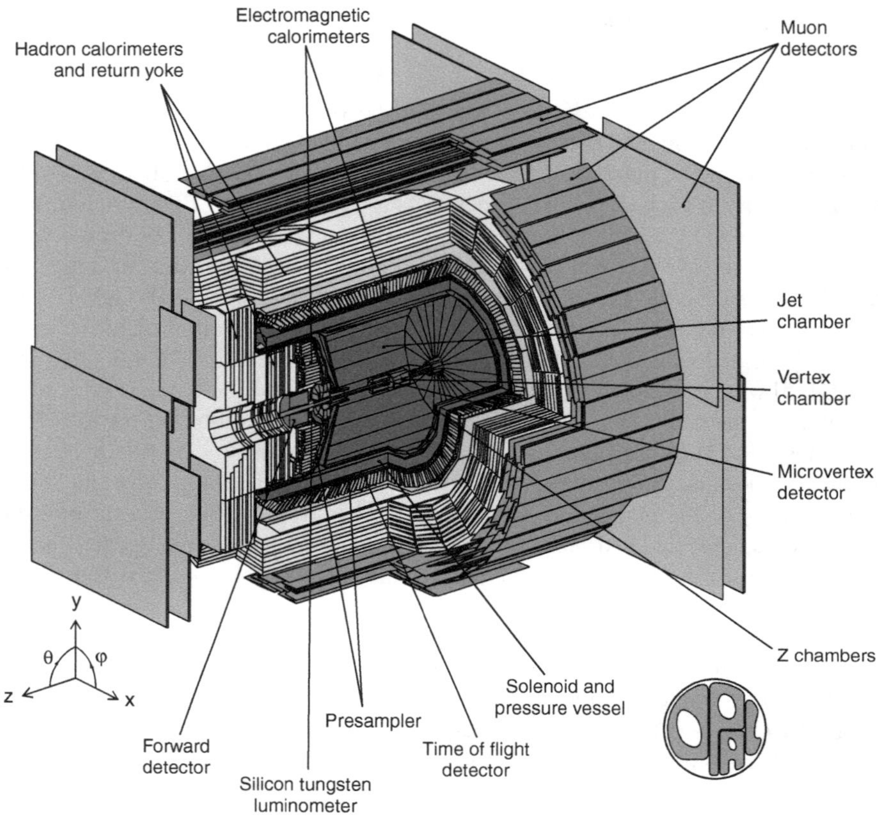

Fig. 1.1 The components of the OPAL detector

When viewed at shorter distance scales, $Z^0 \rightarrow b\bar{b}$ events show a complicated topology illustrated in Fig. 1.3. Following the production and fragmentation of the $\bar{b}b$ quarks, a b-hadron pair (B-$\bar{B}$) move apart back-to-back following the trajectory of the parent quarks. B-hadrons at LEP are produced with average momenta of $p \sim 30\,\text{GeV}$ (a boost of $\gamma\beta \sim 6$) resulting in a typical *decay length* of $\sim 3\,\text{mm}$ before decaying at a *secondary vertex* . In most cases the B will decay to some kind of charm hadron (D) which itself will decay at a *tertiary vertex* point. How vertex detectors were able to reconstruct such an event topology is described further in Chap. 2.

1.3 b-Physics Overview

The LEP b-physics program made measurements spanning many aspects of b-hadron production and decay illustrated schematically in Fig. 1.4.

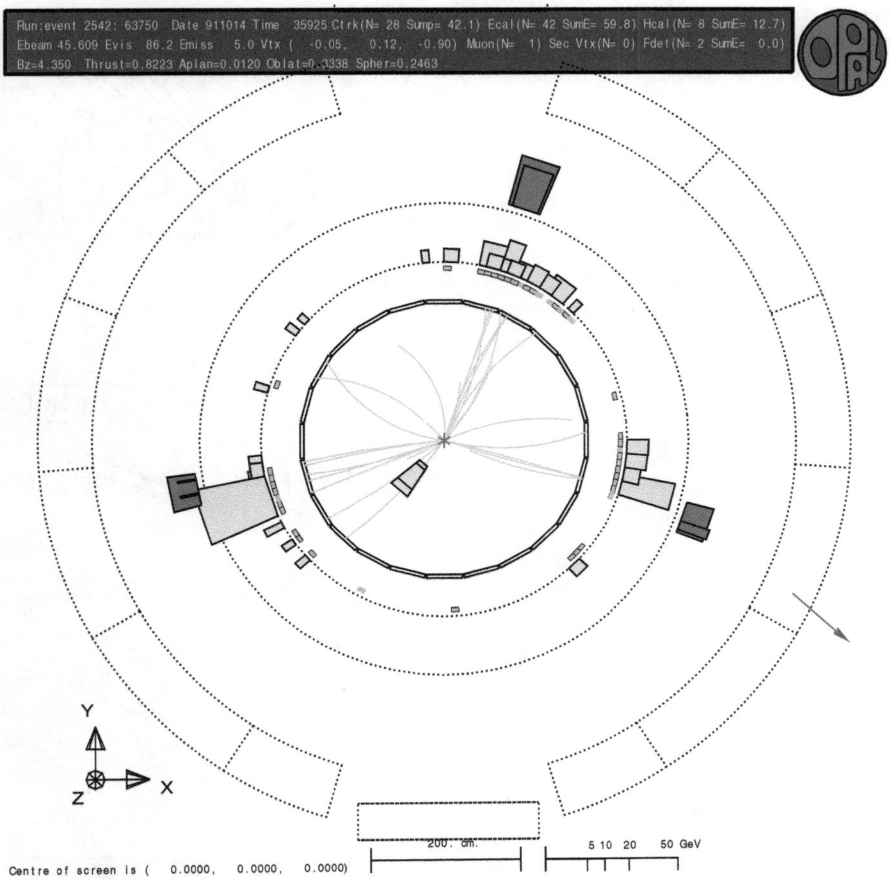

Fig. 1.2 The $r - \phi$ projection of a 3-jet event reconstructed in the OPAL detector

The cross section for $Z^0 \rightarrow b\bar{b}$ at LEP, $\sigma(b\bar{b}) \sim$ few nb, is small relative to b-production at hadron colliders e.g. $\sigma(b\bar{b}) \sim 50 \times 10^3$ nb at the Tevatron. In addition, the large mass of the b-quark means that b-hadrons have literally hundreds of decay channels open to them and so the statistics expected in any one *exclusive* b-decay channel at LEP was prohibitively small. Before LEP start-up these facts were expected to impose a severe limitation on the b-physics program. Effort was therefore focused on developing techniques to use the *inclusive* properties of events for b-physics and details of this work are the focus of our discussion in Chaps. 4, 6 and 7.

In what follows we briefly define theoretical aspects of b-physics that will put into context the experimental measurements. Following the order of Fig. 1.4, we discuss first quantities associated with b-quark production, then b-hadron production and finally b-hadron decay.

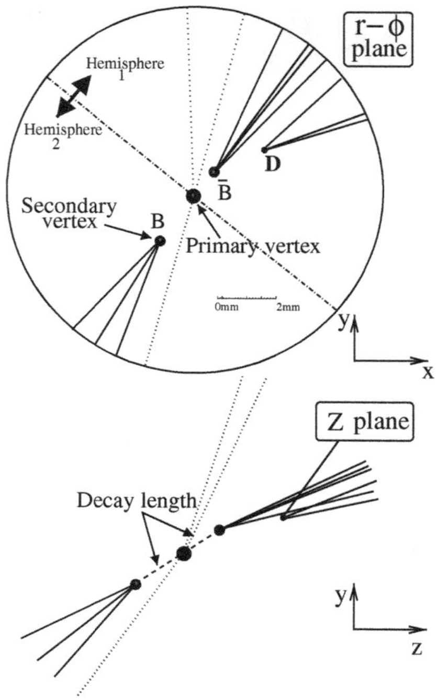

Fig. 1.3 Topology of a typical $Z^0 \rightarrow b\bar{b}$ event

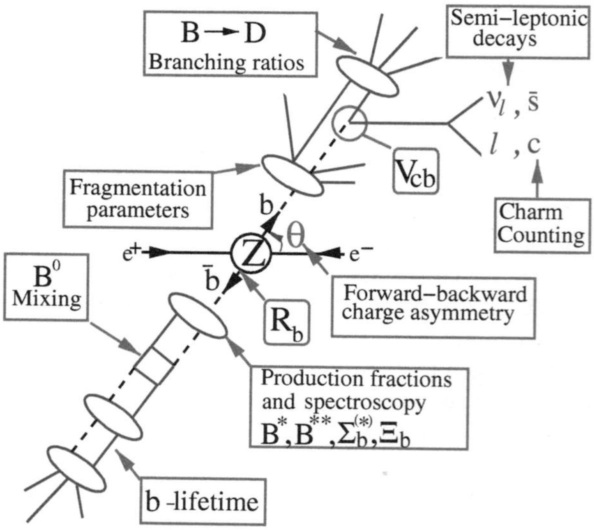

Fig. 1.4 Range of LEP b-physics analysis contributions

1.3.1 b-Quark Production

Close to the Z^0 resonance the Born-level process shown in Fig. 1.5 dominates the fermion production cross section $e^+e^- \to Z^0 \to f\bar{f}$. Working in the Born approximation and assuming massless fermions, the *partial width* for the process $Z^0 \to f\bar{f}$ is given by [7]

$$\Gamma_f = \frac{N_C^f G_F m_Z^3}{6\pi\sqrt{2}} \left[(g_V^f)^2 + (g_A^f)^2 \right] \tag{1.1}$$

N_C^f is a colour factor (3 for quarks, 1 for leptons), G_F is the Fermi coupling constant and g_A^f and g_V^f are the axial and vector couplings to the Z^0:

$$g_A^f = I_3^f \tag{1.2}$$

$$g_V^f = \left(I_3^f - 2Q_f \sin^2\theta_W \right) \tag{1.3}$$

where I_3 is the third component of weak isospin and Q is the electric charge. Moreover, the total cross section at a collision energy $\sqrt{s}$ corresponds to

$$\sigma_f(S) = \frac{12\pi\Gamma_e\Gamma_f}{m_Z^2} \frac{s}{(s - m_Z^2)^2 + \frac{s^2}{m_Z^2}\Gamma_Z^2}. \tag{1.4}$$

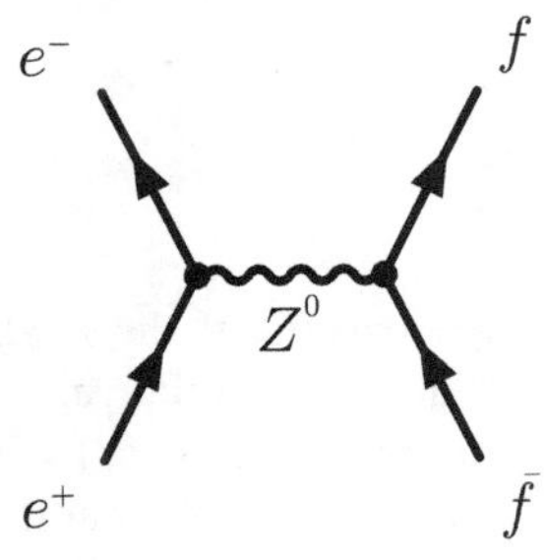

Fig. 1.5 Born level diagram for process $e^+e^- \to Z^0 \to f\bar{f}$

1.3.2 Forward-Backward Asymmetry A_{FB}^b

The differential cross section for $e^+e^- \to Z^0 \to f\bar{f}$ as a function of the polar angle θ between the incoming electron and the outgoing fermion has the following form

$$\frac{d\sigma}{d\cos\theta} = \sigma_f^b \left[\frac{3}{8}(1 + \cos^2\theta) + A_{FB}\cos\theta \right]. \tag{1.5}$$

The forward-backward asymmetry, $A_{\rm FB}$, quantifies the 'amplitude' by which the angular distribution is skewed away from being perfectly symmetric in θ. Experimentally $A_{\rm FB}$ corresponds to the normalised difference between the forward and backward cross section

$$A_{\rm FB} = \frac{\sigma(\cos\theta > 0) - \sigma(\cos\theta < 0)}{\sigma(\cos\theta > 0) + \sigma(\cos\theta < 0)}. \tag{1.6}$$

The asymmetry is directly related to the vector and axial vector couplings of the Z^0 at the electron and fermion vertex depicted in Fig. 1.5 as follows

$$A^{\rm f}{}_{FB} = \frac{3}{4}\frac{2g_V^{\rm e}g_A^{\rm e}}{(g_V^{\rm e})^2 + (g_A^{\rm e})^2}\frac{2g_V^{\rm f}g_A^{\rm f}}{(g_V^{\rm f})^2 + (g_A^{\rm f})^2} \equiv \frac{3}{4}\mathcal{A}_{\rm e}\mathcal{A}_{\rm f} \tag{1.7}$$

Through (1.1), (1.5) and (1.7), measurements of partial widths (or cross sections) and forward-backward asymmetries for the process $e^+e^- \rightarrow q\bar{q}$ give access to the vector and axial couplings of the quarks. It should be emphasised that these expressions are approximations valid only at the Z^0-resonance peak. In general, experiments do not measure directly b-physics parameters at the Z^0-pole but they need to be corrected for radiative effects, γ-exchange and γ-Z interference effects.

With Standard Model values for the couplings and charges it turns out that the forward-backward asymmetry for b-quarks, $A_{\rm FB}^{\rm b}$ is: (a) rather large i.e. around 10% and (b) through the following expression that links the effective weak mixing angle to the couplings,

$$\sin^2\theta_{\rm eff}^{\rm f} = \frac{1}{4|Q_{\rm f}|}\left(1 - \frac{g_V^{\rm f}}{g_A^{\rm f}}\right), \tag{1.8}$$

rather sensitive to the weak mixing angle. For this reason and also because Z^0 decays to b-quarks are relatively easy to identify, $A_{FB}^{\rm b}$ was the most precise measure of the weak mixing angle at LEP.

Finally, we note that if polarised electron beams are available, as was the case at the SLAC SLC collider, one can form a left-right asymmetry

$$A_{\rm LR} = \frac{1}{|\mathcal{P}_{\rm e}|}\frac{(\sigma_{\rm L} - \sigma_{\rm R})}{(\sigma_{\rm L} + \sigma_{\rm R})} \tag{1.9}$$

where $\mathcal{P}_{\rm e}$ is the degree of beam polarisation and L, R denotes use of left or right-handed beams. $A_{\rm LR}$ as defined in (1.9) is actually independent of the final state and the analog of (1.7) for the pole forward-backward asymmetry at LEP is,

$$A_{\rm LR}^0 = \mathcal{A}_{\rm e}. \tag{1.10}$$

The left-right asymmetry therefore directly measures $\mathcal{A}_e$ (instead of the product $\mathcal{A}_e\mathcal{A}_b$ as was the case for A^b_{FB}) and as such provided the single most precise measure of the initial state coupling of the Z^0 to electrons.

The measurement of A^b_{FB} at LEP is is discussed further in Chap. 6. The superb precision finally attained from the LEP collaborations has resulted in a long standing discrepancy with the results from SLC (see Sect. 6.4) and made A^b_{FB} the one electroweak parameter showing a potentially significant discrepancy with the Standard Model.

1.3.3 R_b

Although (1.1) suggests it is measurement of the partial width that is necessary to constrain the electroweak coupling of the b-quark, there are experimental and theoretical reasons why the partial width *ratio*

$$R_b = \frac{\Gamma_b}{\Gamma_{had}} \tag{1.11}$$

is an interesting observable.

Theoretically, many higher order corrections to the calculation of partial widths cancel in such ratios – in particular the correction to the Z^0 propagator due to the unknown Higgs boson mass. Experimentally such ratios have the advantage that several sources of systematic uncertainty common to direct partial width measurements cancel e.g. uncertainties on the luminosity.

R_b is also sensitive to the top quark mass due to the vertex corrections, shown in Fig. 1.6, which are unique to the $Z^0 \rightarrow b\bar{b}$-vertex. Top quark loop contributions such as this enabled the LEP data to predict the mass of the top quark before the Tevatron finally made a direct measurement in 1995 and the agreement of the two methods is a beautiful test of the internal consistency of the Standard Model.[3]

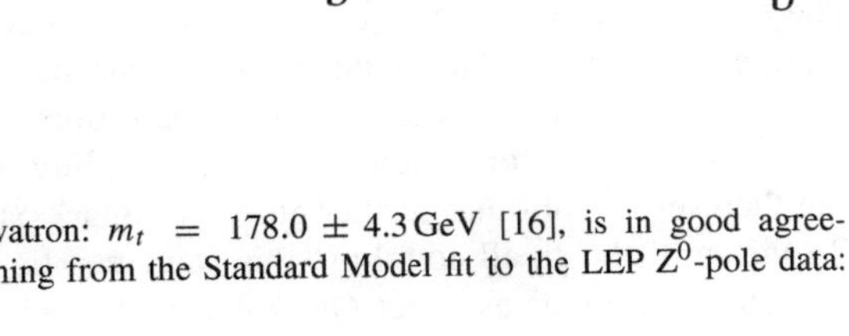

Fig. 1.6 Electroweak corrections to the $Z^0 \rightarrow b\bar{b}$ vertex that involve the top quark

[3] The result from Run I of the Tevatron: $m_t = 178.0 \pm 4.3\,\text{GeV}$ [16], is in good agreement with the indirect constraint coming from the Standard Model fit to the LEP Z^0-pole data: $m_t = 173^{+13}_{-10}\,\text{GeV}$ [17].

The combination of testing a precise theoretical prediction with an experimentally 'clean' measurement makes R_b a good place to search for new physics beyond the Standard Model and there have been instances when the LEP data seemed to indicate that the Standard Model definition of $R_b$was deficient (see Sect. 4.6.1 for further details). The use of R_b as a probe of new physics demanded that the measurements were made with uncertainties of at least 1% which is a level of precision only made possible by the use of double-tag techniques described in Chap. 6.

1.3.4 Heavy Quark Symmetry

Not only did the LEP era trigger rapid development in detector physics and analysis, it also fortuitously coincided with significant advances in theoretical understanding. This was important because the b-quark production and decay weak processes we are interested in, ($Z^0 \rightarrow b\bar{b}$) and ($b \rightarrow cX$), are not what we measure in our experiments: strong interactions bind the $b\bar{b}$ pair from Z^0 decay into B mesons and baryons which subsequently decay $B \rightarrow DX$. In order to use our measurements to draw conclusions about the underlying electroweak physics, the strong interaction contributions must be unfolded. However, perturbative QCD is unable to calculate these bound states and so models must be invoked i.e. fragmentation functions (see Sect. 1.3.5) and quark potential models which naturally introduce uncertainties and can ultimately limit final precision.

More recently an approach has developed based on heavy quark symmetry which has revolutionised the theoretical description of b-hadrons. For details refer to one of the excellent reviews (e.g. [18, 19]) but the basic idea is that in the average b-hadron, the b-quark is so much more massive than the partner u,d,s quarks/anti-quarks and gluons that it can be considered essentially as a static source of colour charge exchanging soft gluons with the light-quark and gluon environment. Under these conditions QCD can be reformulated in terms of an expansion in powers of $\Lambda_{\mathrm{QCD}}/m_b \sim 200\ \mathrm{MeV}/m_b$ which allows model independent predictions to be made with associated uncertainties. This approach, known as Heavy Quark Effective Theory (HQET), has been particularly successful in reducing the theoretical uncertainties in semi-leptonic b-decay measurements (e.g. $|V_{cb}|$ discussed in Sect. 4.3.1).

1.3.5 b-Quark Fragmentation

The *fragmentation* of a $b\bar{b}$ quark pair from Z^0 decay, into jets of particles including the parent b-quarks bound inside b-hadrons, is a process that can be viewed in two stages. The first stage involves the b-quarks radiating hard gluons at scales of $Q^2 >> \Lambda^2_{QCD}$ for which the strong coupling is small $\alpha_s << 1$. These gluons can themselves split into further gluons or quark pairs in a kind of 'parton shower'. By virtue of the small coupling, this stage can be described by perturbative QCD implemented either as exact QCD matrix elements or leading-log parton shower cascade models in software event generators. As the partons separate, the energy scale drops to $\sim \Lambda^2_{QCD}$ and the strong coupling becomes large. We therefore enter a regime where perturbation theory no longer applies. Through the self-interaction of radiated gluons, the colour field energy density between partons builds up to the point where there is sufficient energy to create new quark pairs from the vacuum. This process continues with the result that colourless clusters of quarks and gluons with low internal momentum become bound up together to form hadrons. This 'hadronisation' process represents the second stage of the b-quark fragmentation and because perturbative calculations are not possible, it must be modelled in some way.

Many models of the fragmentation process have been incorporated into simulation packages in the past with varying degrees of success in accurately reproducing the data. In practice these models are implemented via a *fragmentation function* $f_{\mathrm{b}}^{\mathrm{B}}(v)$, parameterised in terms of a kinematical variable v which represents the fraction of the *available* energy that the *b*-hadron receives from the fragmentation process that created it. $f_{\mathrm{b}}^{\mathrm{B}}(v)$ can then be interpreted as the probability density function that a hadron B, containing the original quark b, is produced with a given value v.

A common choice of model for b-quark fragmentation from Z^0 decay, is the 'string' fragmentation scheme of the Lund Group [20]. Here, quark 'popping' from the vacuum is modelled by breaks in a colour flux *string* linking two partons. For an initial b and $\bar{\mathrm{b}}$ quark system recoiling against each other back-to-back, the kinematical variable of the model is defined as

$$z = \frac{(E + p_{||})_{\mathrm{B}}}{(E + p)_{\mathrm{b}}} \tag{1.12}$$

where, $p_{||}$ represents the hadron momentum in the direction of the b-quark and $(E + p)_{\mathrm{b}}$ is the sum of the energy and momentum of the b-quark just before fragmentation begins. From a phenomenological point of view, z is the relevant choice of variable for a parameterisation implemented in an event generator algorithm. However, because z depends explicitly on the properties of the parent b-quark, it is not a quantity that can be directly measured by experiments. What can be measured, via their decay products, are the weakly decaying b-hadrons and so experimental studies of fragmentation from $\mathrm{e^+e^-} \to \mathrm{q\bar{q}}$ data tend to be based on the variable x where,

$$x = \frac{E_{\mathrm{B}}^{\mathrm{weak}}}{E_{\mathrm{b}}} = \frac{E_{\mathrm{B}}^{\mathrm{weak}}}{E_{\mathrm{beam}}} = \frac{2E_B^{\mathrm{weak}}}{\sqrt{s}}. \tag{1.13}$$

E_{B} is the b-hadron energy and $\sqrt{s}$ is the available energy in the centre-of-momentum frame and in the absence of initial state radiation, the quark energy is equal to the electron beam energy.

Because b-quark fragmentation triggers the appearance of b-hadrons, the process affects the kinematics of all b-physics analyses. Measurements that depend on simulation have common fragmentation modelling systematics and this motivates attempts to extract fragmentation parameters in the most model independent way possible – a subject that is discussed further in Sect. 7.2.4.1.

1.3.6 b-Hadron Spectroscopy

Figure 1.7 summarises the expected spectroscopy of b-hadron states that develops at high energy following the production of b-quarks. The numbers are representative branching ratios taken from the DELPHI $Z^0 \to b\bar{b}$ simulation. Of the *primary* b-mesons i.e. those states created directly after the fragmentation phase, about 16% are the weakly decaying pseudoscalar states ($J^P = 0^-$), 49% are vector B*mesons ($J^P = 1^-$) and 35% are orbitally excited states labelled generically as B**. The B** states are potentially an excellent test of the predictions of HQET. HQET demands that the spin of the *b*-quark (S_b) in a hadron is conserved in production and decay processes *independently* of the total angular momentum of the light quark spectator system ($J_q = S_q \oplus L$). Each energy level of the $b\bar{q}$-system thus consists of a pair of states labelled by J_q and the *total spin* $J = J_q + S_b$, i.e. by $J = J_q \pm 1/2$. For the orbitally excited states with $L = 1$, combining

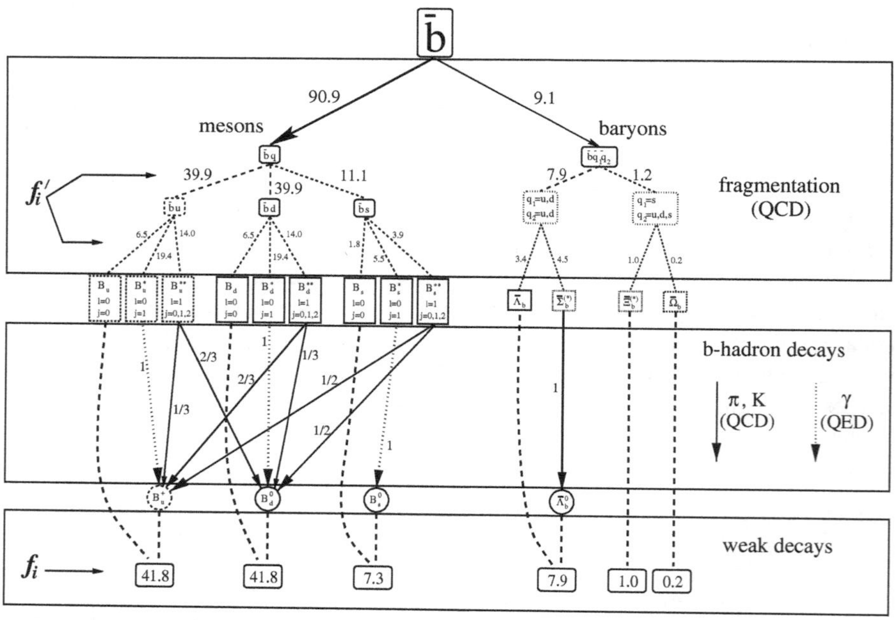

Fig. 1.7 b-Hadron Spectroscopy. The rates of primary hadrons produced in the fragmentation are denoted f' those decay through the weak interaction (indicated by *dashed lines*) by f. Strong and electromagnetic decays are indicated by *solid* and *dotted arrows* together with their expected branching ratios (for strong decays only single pion and kaon transitions considered). From [21]

Table 1.1 Predicted properties of the four $B^{**}_{u,d}$ states

J_q	J^P	B** state	Decay mode	Decay width
1/2	0^+	B_0^*	$B\pi$	Broad ($L_p = 0$, S-wave)
1/2	1^+	B_1^*	$B^*\pi$	Broad ($L_p = 0$, S-wave)
3/2	1^+	B_1	$B^*\pi$	Narrow ($L_p = 2$, D-wave)
3/2	2^+	B_2^*	$B^*\pi, B\pi$	Narrow ($L_p = 2$, D-wave)

$J_q = (1/2, 3/2)$ with $S_b = \pm 1/2$ gives *four* different B** meson states with $J = 0$, 1, 1 and 2. Parity conservation limits the orbital angular momentum between the decay products (L_p) to be even values and demanding that S_b and J_q are separately conserved, leads to the expected decay modes for the four $B^{**}_{u,d}$ states listed in Table 1.1. A similar pattern is expected for the B_s^{**} states although in this case, isospin conservation demands that the only decay channels available are to $B^{(*)}K$ if the B_s^{**} mass is above threshold. The members of the $J_q = 3/2$ doublet are expected to be very narrow compared to typical strong decay widths since only the suppressed $L_p = 2$ (D-wave) transitions are allowed.

Pre-LEP, apart from the Υ resonance states, only the B_u and B_d states had been seen. The LEP data were able to establish the B_s, B^* and Λ_b plus give a strong indication that B**'s were present – discussed in Sect. 7.2.2.1.

1.3.7 b-Hadron Production Fractions

The b-hadron production fractions are defined as the probability of a $\bar{b}$- or b-quark to fragment into the corresponding weakly decaying b-hadron state i.e. $f_{B_i} = BR\left(\bar{b} \to B_i\right) = BR\left(b \to \bar{B}_i\right)$, $B_i = B^+, B_d^0, B_s^0$ or b-baryon. A distinction has to be made between the fractions of b-hadrons produced directly in the fragmentation process (denoted f_i' in Fig.1.7) and the fractions of weakly-decaying states (f_i). The strong decays of orbitally excited ($L = 1$) B** mesons can lead to $f_i' \neq f_i$ e.g. the decays $B_s^{**} \to B^{(*)}K$ results in $f'_{B_s} > f_{B_s}$, $f'_{B_u} < f_{B_u}$and $f'_{B_d} < f_{B_d}$. However, as seen Fig. 1.7, the equality $f'_{B_u} = f'_{B_d}$ is expected to still hold for weakly decay b-hadrons, $f_{B_u} = f_{B_d}$.

Precise knowledge of f_{B_u}, f_{B_d} is desirable since they constitute an important input and systematic error for many b-physics analyses. In addition the fractions are interesting their own right through the insight they give into the start of the fragmentation process. This follows because b-quarks at LEP are almost entirely produced from the decay of the Z^0 with negligible contributions from subsequent fragmentation or gluon splitting $g \to b\bar{b}$.

1.3.8 Weak b-Quark Decay

In the Standard Model with three generations, it is the Cabibbo, Kobayashi and Maskawa (CKM) matrix that rotates the quark mass eigenstates into the weak (or flavour) eigenstates so that the weak charge current becomes,

$$J^\mu = \left(\bar{\mathrm{u}}\ \bar{\mathrm{c}}\ \bar{\mathrm{t}}\right) \frac{\gamma^\mu(1-\gamma^5)}{2} \begin{pmatrix} V_{\mathrm{ud}} & V_{\mathrm{us}} & V_{\mathrm{ub}} \\ V_{\mathrm{cd}} & V_{\mathrm{cs}} & V_{\mathrm{cb}} \\ V_{\mathrm{td}} & V_{\mathrm{ts}} & V_{\mathrm{tb}} \end{pmatrix} \begin{pmatrix} \mathrm{d} \\ \mathrm{s} \\ \mathrm{b} \end{pmatrix} \tag{1.14}$$

The elements of the CKM matrix must be measured, although the unitarity of the matrix means that the values are not independent. Measurements of b-hadron decays at LEP gave information on the following elements:

- $|V_{\mathrm{cb}}|$ from measurements of the mean b-hadron lifetime and studies of $\bar{\mathrm{B}}^0_{\mathrm{d}} \to \mathrm{D}^{*+}\ell^-\bar{\nu}_\ell$ and $\bar{\mathrm{B}}^0_{\mathrm{d}} \to \mathrm{D}^+\ell^-\bar{\nu}_\ell$ decays,
- $|V_{\mathrm{ub}}|$ from (rare) charmless $\mathrm{b} \to \mathrm{u}$ decays,
- $|V_{\mathrm{td}}|$ from $\mathrm{B}^0_{\mathrm{d}} - \bar{\mathrm{B}}^0_{\mathrm{d}}$ mixing and $|V_{\mathrm{ts}}|$ from $\mathrm{B}^0_{\mathrm{s}} - \bar{\mathrm{B}}^0_{\mathrm{s}}$ mixing.

The most important processes responsible for the decay of the ground-state b-mesons are illustrated in Fig. 1.8. The spectator process in which the light quark partner does not contribute is expected to dominate since the exchange and annihilation diagrams are helicity suppressed. Furthermore, the internal spectator process is colour suppressed with respect to the external process (by a factor $\simeq \left(\frac{1}{3}\right)^2$) since the colour of the quarks from the virtual-W must match the colour of the quarks from the parent meson in order to give colourless decay products.

The decay modes of the b-hadrons can be generally classified as either leptonic, semi-leptonic or hadronic where the names are hopefully self-evident. Leptonic decays can only proceed by the annihilation diagram whereas the semi-leptonic

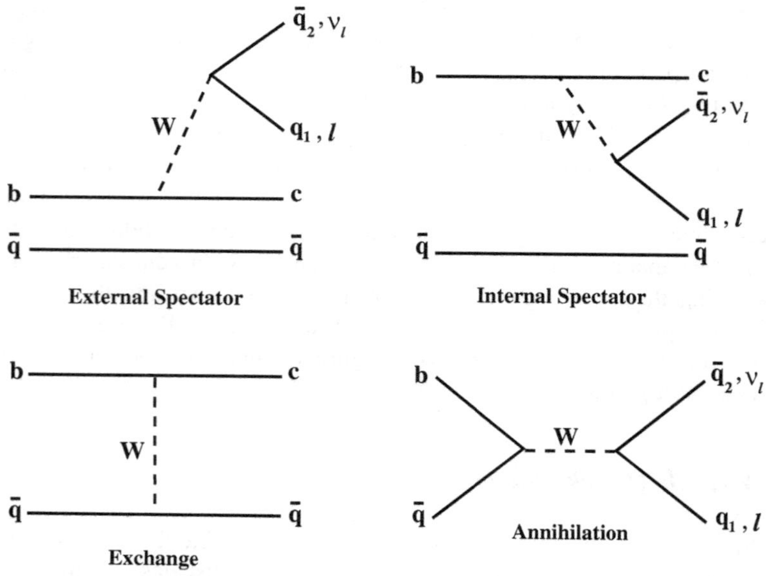

Fig. 1.8 The leading processes contributing to decay of the ground state b-mesons. Penguin and mixing processes and $\mathrm{b} \to \mathrm{u}$ decay also contribute but are suppressed

decays occur by the spectator diagram. Hadronic decays proceed by all three processes and hence dominate the total decay width.

1.3.8.1 b-Hadron Lifetimes

Knowledge of b-hadron lifetimes is fundamental to a b-physics programme not least because experiments typically measure branching fractions whereas theorists predict partial widths. Lifetimes give the connection between them via,

$$\Gamma(\mathrm{B}_i \rightarrow \mathrm{X}_j) = BR(\mathrm{B}_i \rightarrow \mathrm{X}_j) \cdot \frac{1}{\tau(\mathrm{B}_i)}. \tag{1.15}$$

Lifetimes measurements are also very direct tests of the HQET formalism for b-hadron decay since (small) differences in lifetime between the b-hadron species are predicted. In the early days of LEP, before experiments were routinely running silicon vertex detectors, resolutions were sufficient only to measure the mean b-hadron lifetime (τ_{b}) i.e. with no discrimination of b-hadron type. This gave a way of measuring CKM element $|V_{\mathrm{cb}}|$ since, $|V_{\mathrm{cb}}| \propto \tau_{\mathrm{b}}^{-1/2}$. With improved resolutions and techniques the lifetimes of specific b-hadrons have been found to be slightly different making the mean lifetime a redundant quantity – the most precise measurements $|V_{\mathrm{cb}}|$ now come from measurements of the b-quark semi-leptonic branching ratio.

If the spectator diagrams of Fig. 1.8 were the only processes contributing to b-hadron decay, the lifetimes of the different b-hadron species would all be equal. The main effect which causes a difference in lifetime between the mesons is a destructive interference between the internal and external spectator diagrams which can give the same final states for B^+ but not for B_d^0 and B_s^0. This leads to the expected hierarchy: $\tau(\mathrm{B}_\mathrm{d}^0) \sim \tau(\mathrm{B}_\mathrm{s}^0) < \tau(\mathrm{B}^+)$. For b-baryons the situation is rather complex but the expectation is for the lifetimes to be shorter than for the mesons largely because the exchange diagram is no longer helicity suppressed and can even rival the contribution of the spectator process[4]. The same hierarchy is well established in the charm sector but because the lifetime differences scale as $1/m_b^2$ the effects should be much smaller for b-hadrons i.e. on the order of 10% or smaller. Lifetime measurements from the LEP experiments are discussed further in Sect. 7.2.3.1.

1.3.8.2 Semi-leptonic b-Hadron Decays

The total b-hadron decay width is,

$$\begin{aligned}\Gamma_{\mathrm{tot}} = {} & \Gamma(\mathrm{b} \rightarrow \mathrm{c}e\bar{\nu}_e) + \Gamma(\mathrm{b} \rightarrow \mathrm{c}\mu\bar{\nu}_\mu) + \Gamma(\mathrm{b} \rightarrow \mathrm{c}\mu\bar{\nu}_\tau) \\ & + \Gamma(\mathrm{b} \rightarrow \mathrm{c}\bar{\mathrm{u}}\mathrm{d}) + \Gamma(\mathrm{b} \rightarrow \mathrm{c}\bar{\mathrm{c}}\mathrm{s}) + \Gamma(\mathrm{b} \rightarrow \text{no charm})\end{aligned} \tag{1.16}$$

[4] In fact the B_c^+ meson is expected to have a lifetime shorter even than the baryons since both quarks can decay weakly. This has recently been confirmed by the Tevatron experiments but was too rare to be measured at LEP.

where fully leptonic decays are negligible and have been ignored and 'no charm' refers to the small contribution made to the total decay width from rare processes such as $b \to uX$ or $b \to sg(\gamma)$. The semi-leptonic branching ratio is then defined as the ratio of the semi-leptonic partial width to the total decay width of the b-quark, $BR_{sl}^{b} = \Gamma(b \to c\ell\bar{\nu}_{\ell})/\Gamma_{tot}$. In this context, '$\ell$' generally stands for an electron or muon since the tau decay channel is heavily suppressed due to the reduced available phase space. Measurements from the LEP experiments of the semi-leptonic branching ratio produced $BR_{sl}^{b} = 10.59 \pm 0.22$ [22] and measurements based on electrons and muons separately have verified lepton universality in the Standard Model.

The mean number of charm quarks plus anti-quarks produced per b-decay (n_c) is a quantity closely related to b-hadron branching ratios and comparing measurements in (BR_{sl}, n_c)-space to theory predictions provides a stern test. The two quantities are in fact anti-correlated since changes in the semi-leptonic width must induce changes in the hadronic decay width (and hence n_c) in the opposite sense via (1.16). The 'traditional' way to measure n_c is to count up the measured branching ratios of as many exclusively reconstructed c-hadrons as possible

$$n_c = BR(b \to DX) + 2BR(b \to (c\bar{c})X), \tag{1.17}$$

where $D \equiv D^+, D^0, D_s^+$ and charm baryons. Corrections must be made from theory for any unmeasured baryons and charmonium ($c\bar{c}$) states. LEP measurements of n_c by this method have been made by ALEPH [23], DELPHI [24] and OPAL [25] based on their samples of exclusively reconstructed charm states but the results are ultimately limited by the uncertainties on the c-hadron decay branching ratios reported in Table 1.2. Better precision on n_c was possible from measurements of the rate for no-charm and two-charm b-decays through the following relationship,

$$\begin{aligned} n_c = {} & [1 - BR(b \to \text{no} - \text{charm})] \\ & + BR(b \to c\bar{c}s) + BR(b \to (c\bar{c})X). \end{aligned} \tag{1.18}$$

In practice, measurements of the no-charm rate include the 'hidden-charm' contribution from the charmonium events i.e.

$$BR^{\text{meas.}}(b \to \text{no} - \text{charm}) = BR(b \to \text{no} - \text{charm}) + BR(b \to (c\bar{c})X). \tag{1.19}$$

Substituting for $BR(b \to \text{no} - \text{charm})$ from above into 1.18 gives,

$$\begin{aligned} n_c = {} & 1 - BR^{\text{meas.}}(b \to \text{no} - \text{charm}) \\ & + BR(b \to c\bar{c}s) + 2BR(b \to (c\bar{c})X). \end{aligned} \tag{1.20}$$

The contribution of LEP to the (BR_{sl}, n_c) debate is discussed further in Sect. 7.3.1.

1.3.8.3 B^0 Oscillations

In the same way that neutral kaon states mix, the B^0 and $\bar{B}^0$ meson states that undergo weak interactions (the 'flavour' states) are expected to be mixtures of the mass eigenstates. Neglecting the complication of CP-violation which is anyway small compared to the mixing effects, the flavour states can be written as

$$B^0 = \frac{1}{\sqrt{2}}\left(B^0_L + B^0_H\right) \; ; \; \bar{B}^0 = \frac{1}{\sqrt{2}}\left(B^0_L - B^0_H\right) \tag{1.21}$$

where B^0_L, B^0_H are the 'Light' and 'Heavy' mass eigenstates. The box diagram processes shown in Fig 1.9 are responsible for the mass difference $\Delta m = (m_H - m_L)$ which introduces a time dependent phase shift between mass eigenstate wave functions. A consequence of this are the 'oscillation' probabilities for an initial B^0 state to decay either as a $\bar{B}^0$ (mixed) or as a B^0 (unmixed),

$$P_{\text{mix}} = \frac{1}{2\tau} e^{-t/\tau}(1 - \cos \Delta m t) \tag{1.22}$$

$$P_{\text{unmix}} = \frac{1}{2\tau} e^{-t/\tau}(1 + \cos \Delta m t) \tag{1.23}$$

where τ is the B^0 mean lifetime. These probabilities are nothing more than an envelope of standard exponential decay with mean lifetime τ, modulated by an oscillation of frequency Δm. Figure 1.9 plots these mixed and un-mixed probabilities for the case of a small oscillation frequency, where the state decays before it has time to mix, and a high frequency where many oscillations occur on average before decay. Measurements made pre-LEP and early results from LEP, were unable to resolve the oscillations in time and measured only the time integrated probability

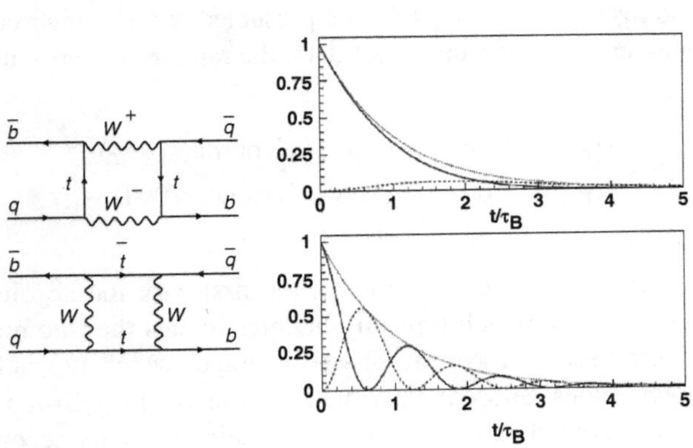

Fig. 1.9 (*left*) The 2nd-order 'box' diagram processes that generate B^0 oscillations (from [22]); (*right*) the time evolution of B^0 oscillations at two different frequencies showing the unmixed (*solid*) and mixed (*dashed*) contributions together with their sum (from [26])

$$\chi = \frac{P_{\text{mix}}}{P_{\text{mix}} + P_{\text{unmix}}} = \frac{x^2}{2(1+x^2)} \tag{1.24}$$

where $x = \Delta m\tau$ i.e. the ratio of the lifetime to the oscillation time. The parameter χ also does not discriminate between the two neutral flavour states possible: B^0_d and B^0_s. To measure the oscillations in time (and hence Δm) for the flavour states separately is a tall order experimentally and, as we have already noted, was not thought possible before the start of the LEP b-programme. The measurement demands knowledge of the B^0-flavour at both the production and decay time in addition to the decay time itself. Following the development of efficient flavour tagging techniques coupled with a favourably small oscillation frequency, oscillations of the B^0_d were eventually measured at LEP with precision of a few percent. The B^0_s however turned out to have a much higher oscillation frequency making the oscillation time so short that it proved impossible for the LEP experiments to resolve. The measurement of B^0_s has had to wait for a recent result from CDF with a huge sample of exclusive decays and applying flavour tagging techniques first developed at LEP - see Sect. 7.3 and Chap. 8.

1.4 Combining b-Physics Results

For many quantities, the precision possible from a single LEP experiment was not sufficient to probe the predictions of the Standard Model. It was therefore of crucial importance that joint structures were put in place across the collaborations whose remit was to combine in a statistically rigorous way measurements from the Z^0 data.

This is difficult work since the resulting averages can be significantly different depending on the way statistical errors are handled and on the assumptions made concerning the correlated systematic uncertainties. This is well illustrated by the (in)famous world average of the B^0_s lifetime presented at the Winter conferences in 1994. Based on essentially the same input data, the rapporteurs presented,

$$\langle \tau_{B_s} \rangle = 1.38 \pm 0.17\,\text{ps (LaThuille, 1994)}$$
$$\langle \tau_{B_s} \rangle = 1.66 \pm 0.22\,\text{ps (Moriond, 1994).}$$

The difference derived from the fact that in the first case the absolute error was used to weight the values, which typically underestimates the true mean, whereas relative errors were used to form the second average, which typically biases the true mean to larger values. Subsequently the Lifetime Working Group found that if enough values are averaged and correlated systematic errors are accounted for, the two approaches were found to converge on the correct mean value to within about 1%. The (small) difference between the two is now assigned as a systematic on the averaging procedure itself.

1.4.1 Averaging of Electroweak Heavy Flavour Quantities

Heavy flavour electroweak parameters play an important role in testing the consistency of experimental measurements with the predictions of the electroweak theory. Almost all the b-physics quantities have been measured by several methods and by several different experiments. All measurements require some input from simulation which are subject to uncertainties: (i) related to the modelling of detector response, which are uncorrelated between different experiments, and (ii) arising from limited knowledge of the generated physics processes, which are correlated between experiments.

To facilitate the treatment of common systematic errors in averages of heavy flavour electroweak parameters, the LEP collaborations and SLD agreed on a common set of simulation input parameters and their errors within the framework of the Electroweak Heavy Flavour Working Group - a sub-group of the LEP Electroweak Working Group (EWWG). Some of the more important input parameter settings are listed in Table 1.2.

Table 1.2 Common simulation inputs for heavy flavour electroweak measurements. See [27] for full details

Quantity	Value
Beam energy fraction for b-quarks	$\langle x_b \rangle = 0.702 \pm 0.008$
Average b-hadron lifetime	$\tau_b = 1.576 \pm 0.016$ ps
b-quarks from gluons	$P(g \to b\bar{b}) = 0.254 \pm 0.051\%$
b-decay charged multiplicity	$\langle n_b^{ch} \rangle = 4.955 \pm 0.062$
Beam energy fraction for c-quarks	$\langle x_c \rangle = 0.484 \pm 0.008$
c-quarks from gluons	$P(g \to c\bar{c}) = 2.96 \pm 0.38$
D-topological branching ratios	$f_i(D^0, D^+, D_s)$ $i = 0, .., 6$ charged particles, set to the results from MARKIII [28]
D^0 lifetime	$\tau\left(D^0\right) = 0.415 \pm 0.004$ ps
D^+ lifetime	$\tau\left(D^+\right) = 1.057 \pm 0.015$ ps
D_s lifetime	$\tau\left(D_s\right) = 0.467 \pm 0.017$ ps
Λ_c lifetime	$\tau\left(D_s\right) = 0.206 \pm 0.012$ ps
D^0 branching ratio	$BR\left(D^0 \to K^-\pi^+\right) = 0.0385 \pm 0.0009$
D^+ branching ratio	$BR\left(D^+ \to K^-\pi^+\pi^+\right) = 0.090 \pm 0.006$
D_s branching ratio	$BR\left(D_s \to \phi\pi^+\right) = 0.036 \pm 0.009$
	$BR\left(D_s \to \bar{K}^{*0}K^+\right) / \left(D_s \to \phi\pi^+\right) = 0.92 \pm 0.09$
Λ_c branching ratio	$BR\left(\Lambda_c \to pK^-\pi^+\right) = 0.050 \pm 0.013$

The averaging of electroweak measurements is performed as a χ^2 fit. The input covariance matrix is formed from a table supplied by each analysis giving the result and the error breakdown, including entries for all agreed common sources of error listed in Table 1.2. The heavy flavour electroweak quantities to be combined are essentially the normalised partial widths R_b, R_c and the asymmetries i.e. forward-backward A_{FB}^b, A_{FB}^c from the LEP experiments and $\mathcal{A}_b$, $\mathcal{A}_c$ obtained from the the left-right asymmetries at SLD. If more than one quantity is measured in the same analysis the correlations between them are also supplied by the experiment. If a result depends on another fit parameter which is not measured in the same analysis, the experiment supplies the dependence of the result on this parameter so that it can be varied during the fit.

Further details of the combination methodology and the current status of the results are available at: `http://lepewwg.web.cern.ch/LEPEWWG/` and the references therein.

1.4.2 Averaging of Non-electroweak Heavy Flavour Quantities

Working groups were established in the following areas to address the problem of correctly combining the (non-electroweak) b-physics measurements from the LEP collaborations and from around the world:

- b-hadron lifetimes,
- B_d^0 and B_s^0 oscillations,
- $|V_{cb}|$ and $|V_{ub}|$,
- CKM unitarity triangle parameters.

The effort was originally coordinated by the Heavy Flavour Steering Group and details of the averaging procedures adopted are given in [8]. Since 2002 this task has been under the direction of the Heavy Flavour Averaging Group (HFAG) .

Before a combination of results can occur, each individual measurement is first adjusted to the same set of physics inputs used for the electroweak heavy flavour fit (Table 1.2) or if the parameter is part of the electroweak combination described above, the fit result is taken. These adjustments are performed if (and only if) a systematic uncertainty associated to a given physics parameter has been quoted by the experiment. The adjustment procedure affects both the central value of the measurement (by an amount proportional to the quoted systematic uncertainty) and the relevant systematic uncertainty. The large number and range of different analyses makes it difficult to organise a common treatment of systematic effects and the evaluation of correlations. Therefore no global fit of quantities similar to the electroweak averaging scheme is made and instead a combination method for each measurement is defined. Since the quantities averaged by the HFAG tend to be of lower precision than the electroweak quantities, this does not currently limit the final precisions attained. The most recent averages are made available at: `http://www.slac.stanford.edu/xorg/hfag/index.html`.

References

1. S. Myers: *The LEP collider, from design to approval and commissioning*, sixth John Adams Memorial Lecture, CERN 91-08, (1991)
2. B. Richter: *Very high energy electron-positron colliding beams for the study of the weak interaction*, CERN/ISR-LTD/76-9, (1976)
3. MAC Collab.: Phys. Rev. Lett. **51**, 1022 (1983)

4. MARKII Collab.: Phys. Rev. Lett. **51**, 1316
5. J.H. Kühn, P.M. Zerwas et al.: Heavy Flavours. In: *Z Physics at LEP 1*, vol 1, G. Altarelli, R. Kleiss and C. Verzegnassi (eds.) (CERN 89-08) pp. 267–372
6. M.G. Green, S.L. Lloyd, P.N. Ratoff, D.R. Ward: *Electron-Positron Physics at the Z*, (IOP, 1989)
7. K. Mönig: Rep. Prog. Phys. **61**, 999 (1998)
8. ALEPH, CDF, DELPHI, L3, OPAL, SLD: Combined results on b-hadron production rates and decay properties. CERN-EP/2001-050
9. P. Roudeau: Ten years of b-physics at LEP and elsewhere. In: *29th International Meeting on Fundamental Physics IMFP XXIX*, Sitges, Barcelona (2001)
10. ALEPH Collab., D. Decamp et al.: Nucl. Instr. Meth. A **294**, 121 (1990)
11. ALEPH Collab., D. Buskulic et al.: Nucl. Instr. Meth. A **360**, 481 (1995)
12. DELPHI Collab., P. Aarnio et al.: Nucl. Instr. Meth. A **303**, 233 (1991)
13. DELPHI Collab., P. Abreu et al.: Nucl. Instr. Meth. A **378**, 57 (1996)
14. L3 Collab., O. Adriani et al.: Phys. Rep. **236**, 1 (1993)
15. OPAL Collab., K. Ahmet et al.: Nucl. Instr. Meth. A **305**, 275 (1991)
16. CDF, D0 and the Tevatron Electroweak Working Group: Combination of CDF and D0 results on the top quark mass. hep-ex/040410, 2004
17. ALEPH, DELPHI, L3, OPAL, SLD, LEP Electroweak Working Group, SLD Electroweak Group and SLD Heavy Flavour Group: Phys. Rep. **427**, 257 (2006)
18. M. Neubert: Introduction to B Physics. In:*Lectures presented at the Trieste Summer School in Particle Physics (Part II)*, hep-ph/0001334 (2000)
19. A. F. Falk: The heavy quark expansion of QCD. In:*Proc. XXIVth SLAC Summer Institute on Particle Physics*, Stanford (1996)
20. B. Anderson, G. Gustafson, G. Ingelman, T. Sjöstrand: Phys. Rep. **97**, 31 (1983)
21. DELPHI Collab., J. Abdallah et al.: Phys. Lett. B **576**, 29 (2003)
22. K. Hagiwara et al.: Phys. Rev. D **66**, 010001 (2002) and 2003 off-year partial update for the 2004 edition available on the PDG WWW pages (URL: http://pdg.lbl.gov/)
23. ALEPH Collab., D. Buskulic et al.: Phys. Lett. B **388**, 648 (1996)
24. DELPHI Collab., P. Abreu et al.: Eur. Phys. J. **12**, 225 (2000)
25. OPAL Collab., G. Alexander et al.: Z. Phys. C **72**, 1 (1996)
26. M. Paulini: Int. J. Mod. Phys. A **14**, 2791 (1999)
27. LEP/SLD Heavy Flavour Working Group: Final input parameters for the LEP/SLD heavy flavour analyses. LEPHF/2001-01
28. MARKIII Collab., D. Coffman et al.: Phys. Lett. B **263**, 135–140 (1991)

Chapter 2
Silicon Vertex Detectors and Particle Identification

The ability to reconstruct the decay vertex of a b-hadron and simultaneously to know the type of particle associated with this vertex were the essential tools of inclusive b-physics at LEP. The demands of an ever more ambitious b-physics program at LEP (and also at the SLC) were the driving force behind the rapid development of silicon strip detectors capable of operating in the challenging environment of a collider. At the same time 'traditional' particle identification techniques such as ionisation energy loss $\mathrm{d}E/\mathrm{d}x$ were being supplemented with radical new methods such as Ring Imaging Čerenkov Counters in order to give a particle identification capability at a level and scale never before attempted.

2.1 Silicon-Strip Vertex Detectors

Silicon strip detectors are able to detect the trajectory of charged particles so precisely that tracks extrapolated through the silicon hits and into the radius of the LEP beampipe, were able to resolve the quite complicated decay vertex topology of b-hadron decays. This is well illustrated by Fig. 2.1 which shows a candidate $Z^0 \rightarrow b\bar{b}$ event reconstructed by the ALEPH tracking and calorimetry sub-detectors. Two further views provide a close-up of how tracks reconstructed in the outer tracking volumes match with hits in the silicon and the lower view zooms in on the interaction region showing how the extrapolated tracks meet at vertex points. The excellent precision separates clearly the primary Z^0 decay vertex from three decay vertices, each associated with an error ellipse, which are likely to have originated from the decay of a B or $\bar{B}$ producing a D-meson which itself decays some distance from the parent B into three tracks. Reconstruction of this type of topology would have been impossible with the relatively poor extrapolation resolution provided by the 'standard' tracking detectors of the LEP experiments e.g. drift chambers or TPC's.

The physics of silicon-strip detectors has been extensively reported in the literature and will not be repeated here (for an excellent review, see e.g. [1]). Instead we will concentrate on those aspects of detector development which enabled the LEP b-physics program to function.

G.J. Barker, *b-Quark Physics with the LEP Collider*, STMP 236, 23–35,
DOI 10.1007/978-3-642-05279-8_2, © Springer-Verlag Berlin Heidelberg 2010

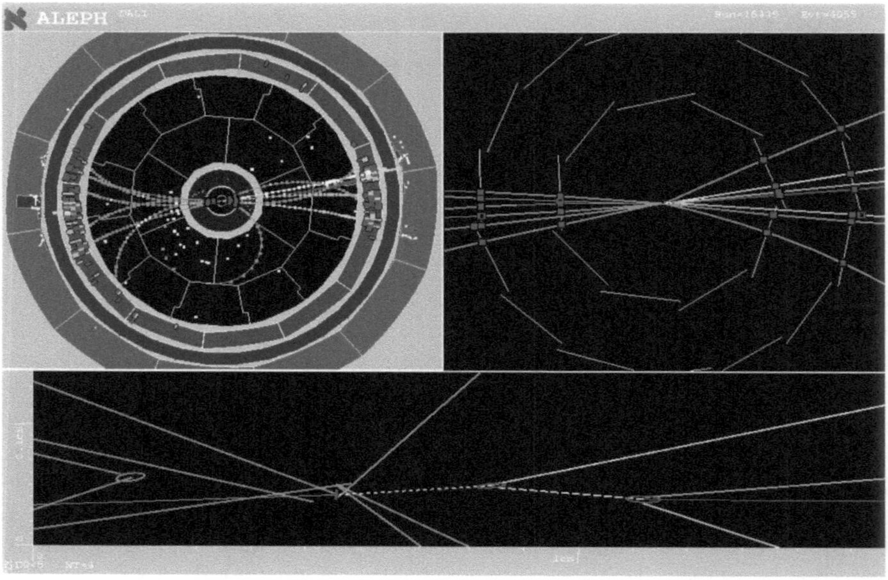

Fig. 2.1 A candidate $Z^0 \rightarrow b\bar{b}$ event reconstructed by the ALEPH silicon vertex detector

High energy physics experiments first used semiconductor devices in the 1970s to measure particle energies but it was not until the advent of the planar production technique of silicon *p-n* junction diodes by Kemmer [2] in 1980, that such detectors could be used also to measure particle trajectories. By segmenting one side of the junction into thin strips, particle positions are located according to which strip(s) collected a charge signal. The basic idea behind these silicon-strip detectors is illustrated in Fig. 2.2. The spatial resolution is determined by the strip spacing or 'pitch', which would ideally be smaller than the width of the charge distribution released from the passage of an ionising particle i.e. $\mathcal{O}10$ μm. In practice strip pitches of between 20 and 50 μm can be achieved but spatial resolutions better than the geometrical limit of $pitch/\sqrt{12}$ can be reached by reconstructing offline the 'centre of gravity' of the charge distribution. The planar technique also allowed the possibility of simultaneously processing many silicon wafers so increasing the production yield and lowering the associated costs.

Silicon strip detectors were used successfully in fixed target experiments in the early 1980s to study charmed particles but before their use was possible in a collider experiment a number of developments were needed.

Fixed target experiments had the advantage of essentially unlimited space outside of the beamline and so they were able to use large fan-outs connecting the small strip pitch of the detectors to readout electronics where the channel pitch could be 2–3 orders of magnitude larger. This configuration would be impossible in a collider experiment where as much as possible of the space around the interaction should be instrumented with active detectors and the amount of passive material must be kept

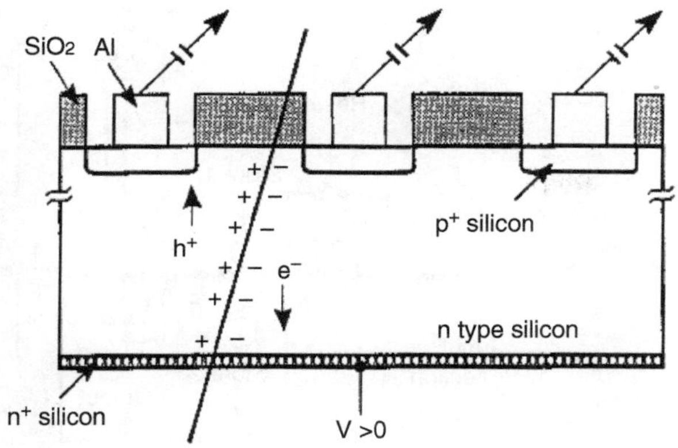

Fig. 2.2 Schematic of a single-sided silicon-strip detector

to a minimum in order to limit the effects of multiple scattering on particle trajectories. It therefore became clear at an early stage that the density of readout channels would need to be much higher than was necessary in fixed target experiments and that detector engineering designs would be severely tested by space limitations and multiple scattering limits. Custom designed, Very Large Scale Integration (VLSI) ASIC's were the technological breakthrough necessary to address these problems. Sub-5 μm feature processing enabled up to 128 readout channels to be crammed into an area of roughly 6 × 6 mm. The first example was the Multiplex chip , based on 5 μm NMOS technology, designed to readout the silicon strip detectors of the MARK II detector. The basic design of each channel, shown in Fig. 2.3, consists of a charge sensitive amplifier followed by two identical capacitor circuits one of which samples the charge before the arrival of the signal and the other after the signal has arrived. The voltage difference between the two is then proportional to the signal charge without the effects of noise (at least for noise with time variations much longer than the period between the two sampling times). This signal is stored and multiplexed onto a serial bus. The LEP experiments subsequently introduced their own variations on this original design which generally had much lower power consumption realised in 5 μm (or smaller) CMOS technology.

In collider-mode a significant fraction of particles cross the detector planes at low incidence angles which spreads the charge deposited over a number of readout strips. This reduces the size of particle signals and means that signal/noise performance of the detectors was much more of an issue for the collider experiments than was the case with fixed-target geometries. The main source of noise in silicon-strip detectors is often due to the capacitance of the strip being read-out, both to the back plane and to neighbouring strips. This causes signal loss and acts as a load capacitance to the channel amplifier which gives electronics noise. This can be minimised by making the coupling capacitance to the read-out electronics high and the inter-strip capacitance as low as possible. The bias resistor, through which the depletion voltage for the silicon substrate is supplied, is another source of noise which can

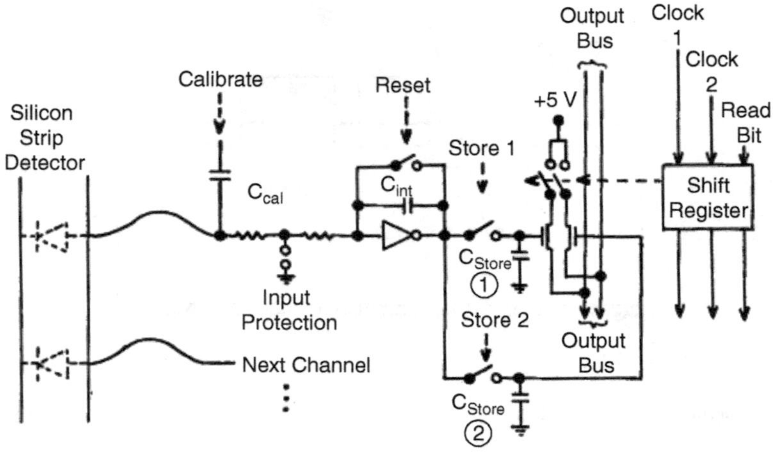

Fig. 2.3 Circuit layout of one channel of the MARK II Multiplex chip. From [3]

be kept under control by making the resistance as large as possible. Detectors were developed that incorporated coupling capacitors (via SiO_2 layers) and poly-silicon bias resistors (of a few$\times$ MΩ) integrated in the detector silicon substrate itself. This type of integration freed-up much needed space in the multiplex-chip designs and helped to keep the amount of material through which particles must pass, down to a minimum. The material budget is particularly important because of the multiple Coulomb scattering it introduces and which limits the spatial resolution attainable for low momentum tracks. The integration of more and more of the ancillary supply and readout electronics onto the silicon wafer was a development that began with the LEP detector designs and continues today in designs for the next generation of detector.

An early development goal of the LEP vertex detector groups was to design a strip detector that could provide 3D spatial points. This effort was pioneered by the ALEPH collaboration who developed double-sided devices where the ohmic contact (or n)-side of the device pictured in Fig. 2.2, was also segmented into strips running in the perpendicular direction to the p-side strips. These devices are complicated to manufacture and early production batches were of varying quality. Because of this OPAL preferred the solution of sticking two single-sided devices, with perpendicular strip orientations, back-to-back. Double-sided devices do however provide the most satisfactory route to 3D reconstruction since the amount of material is essentially no more than for single-sided devices and the correlation between pulse heights collected on the p- and n-sides (so called, Landau correlations) can be used to match signals from both sides with no ambiguity. For these reasons double-sided development at LEP continued and in addition to ALEPH, double-sided sensors were subsequently installed by DELPHI and L3. A comparison of the silicon detectors of the LEP experiments is given in Table 2.1.

A very active area of development concerning double-sided devices, was how to read-out the signals from the n-side strips which are oriented perpendicular to

Table 2.1 Comparison of the silicon vertex detectors of the four LEP experiments in their final configurations

	ALEPH	DELPHI	L3	OPAL
Reference	[4]	[5]	[6]	[7]
No. layers	2	3	2	2
$\langle r_1 \rangle$ (cm)	6.3	6.3	6.1	6.2
$\langle r_2 \rangle$ (cm)	–	9.0	–	–
$\langle r_3 \rangle$ (cm)	10.7	10.9	7.8	7.7
No. detectors	96	288	96	75
Silicon area (m^2)	0.25	0.42	0.30	0.15
Readout-chip	CAMEX64A (3.5 μm CMOS)	MX3/MX6 (3 μm CMOS)	SVXD (3 μm CMOS)	MX5/MX7 (1.5 μm CMOS)
Signal/noise	17	13–16	–	22

the beam line. The simplest scheme, implemented by ALEPH, is to mount read-out electronics directly at the end of the strips which has the disadvantage of introducing extra material. A preferable scheme is to read out both the p- and n-side strips at the ends of sensor modules away from the highest density of traversing particles. OPAL and L3 achieved this by routing the n-side signals to the end of modules by bonding to conducting lines laid down on glass (OPAL) and capton (L3) plates stuck directly onto the sensors. The most elaborate solution to this problem however, came from DELPHI who developed double-sided detectors with an extra, integral, metal layer to re-route the signals. These detectors, Fig. 2.4, represented the state-of-the-art in silicon strip detector design for collider experiments and presented many fabrication challenges for device manufacturers. Worth mention here also is the novel design choice of DELPHI to join neighbouring double-sided detectors in a 'flipped' configuration as shown in Fig. 2.5. Due to the biasing scheme chosen for the strip-devices, the readout lines on the p and n-sides were at the same potential and it was therefore possible to bond the n-side readout lines of one detector to the p-side readout lines of the neighbouring detector. This arrangement was found to have a number of advantages [5] including an equalisation of capacitance noise from both sides, a signal polarity which tags cleanly which detector in the module produced the signal and some benefits for the alignment procedure which is discussed in Chap.3.

In parallel with developments in silicon-strip sensor design and read-out electronics, to make a multi-layer detector at the heart of a collider experiment presented many challenges for mechanical support structures, cable routing, multiplex-chip cooling etc. Spatial constraints were the main enemy and were particularly severe in the case of L3 and OPAL. Silicon detectors did not form part of the original Letters of Intent for these two experiments and were only possible after 1991 when the LEP beampipe radius was reduced from 8.5 to 5.5cm. OPAL installed their silicon vertex detector in 1991 just one year after the project's approval and L3 followed in 1993 2 years after approval. These rapid installation schedules pay testament to how rapidly all aspects of silicon vertex detector construction advanced in the early years of LEP. By 1994 all of the LEP experiments had working devices providing an intrinsic resolution of sub-10 μm in both the $R-\phi$ and $R-z$ planes. In order

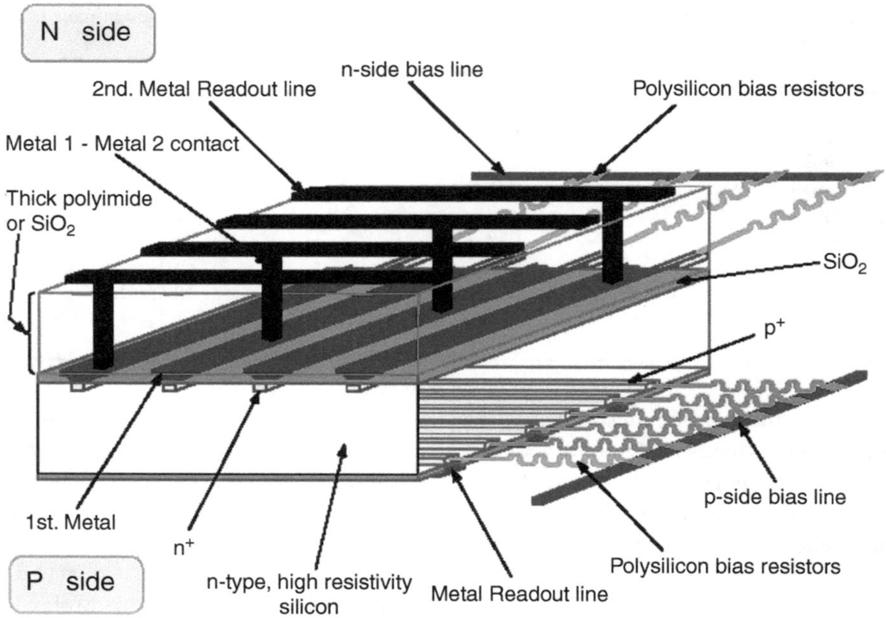

Fig. 2.4 Schematic of the DELPHI double-sided devices with two metal layers for same-end read-out. From [5]

to best exploit this resolution for physics required a careful alignment of the silicon detector modules and the optimisation of off-line reconstruction algorithms. These matters are discussed further in Chap. 3.

One further round of major modification to the LEP vertex detectors was triggered by the onset of the LEP 2 era where beam energy reached the 200 GeV

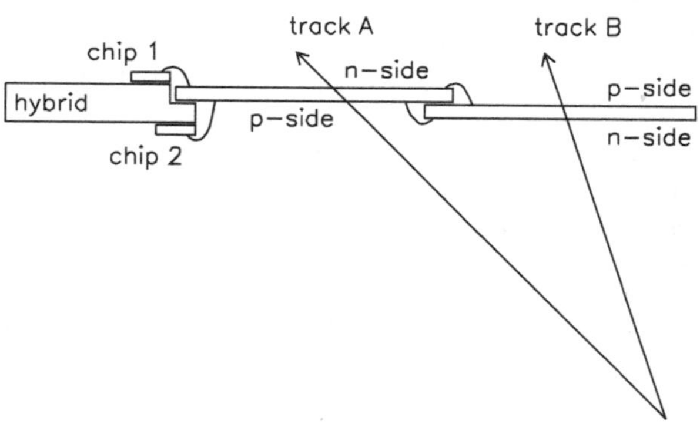

Fig. 2.5 The DELPHI flipped module concept for double-sided devices. Chip 1 will register a negative (positive) signal from tracks A (B), while chip 2 will register a positive (negative) signal from track A (B)

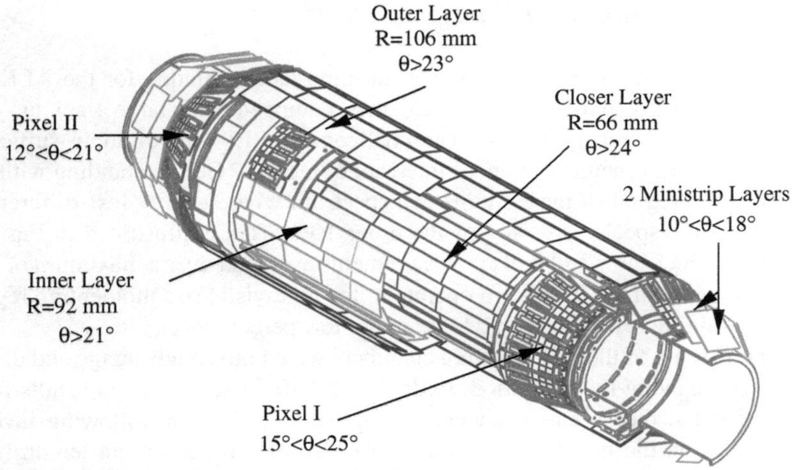

Fig. 2.6 Layout of the DELPHI silicon tracker. From [8]

threshold and higher. In this environment the ability to track particles at low polar angles became important and the experiments responded by extending the solid angle covered by the silicon detectors. The most elaborate upgrade was the DELPHI Silicon Tracker [8] completed in 1997, which added silicon mini-strip and pixel devices in the forward region. The Tracker, pictured in Fig. 2.6, represents the pinnacle of silicon vertex detector development at LEP and remained, until the turn-on of the LHC, the largest collider silicon detector ever constructed.

2.2 Particle Identification

The LEP experiments were multi-purpose devices well equipped to detect and separate the stable products of e^+e^- collisions: photons, electrons, muons, pions, kaons and protons. A broad range of, mainly well-established, techniques based on detecting ionisation and excitation energy loss and time-of-flight were used. Electromagnetic calorimeters reconstructed electron and photon-induced showers and hadron calorimeters performed the same function for hadronic showers. Muons could be identified by associating tracks with hits in the muon chambers around the very outside of the detectors.

In this section we will briefly review aspects of particle identification at LEP which were new or novel in some way and which played an important role in the b-physics program. This work in particular led to advances in how to combine particle identification information in an optimal way and this is discussed further in Chap. 3.

2.2.1 Specific Ionisation Energy Loss, dE/dx

Sampling the ionisation energy loss was an important technique for the ALEPH, DELPHI and OPAL experiments which were equipped with large tracking gas-chambers (diameter up to 3.6 m). The challenge at LEP was to attain sufficient resolution in the momentum range of interest i.e. $p > 1$–2 GeV, coinciding with the relativistic rise region of the Bethe-Bloch spectrum where energy loss differences between particle species are only of the order 10%. This is illustrated in Fig. 2.7 from OPAL which shows $\langle dE/dx \rangle$ measurements averaged over a maximum of 159 samples along the track length. To operate in the relativistic rise momentum region demands attaining resolutions on dE/dx at the few percent level.

The large scale of the LEP tracking chambers were both an advantage and disadvantage to achieving the resolution needed on dE/dx. The advantage results from the fact that the measurement uncertainty, $\sigma(dE/dx)$, has the following inverse dependencies on the number of samples (N) and effective sampling length ($l =$ thickness $\times$ pressure): $\sigma(dE/dx) \propto N^{[-0.43,-0.47]}$ and $\sigma(dE/dx) \propto l^{[-0.32,-0.36]}$ respectively [9]. The disadvantage results from the practical difficulties associated with operating a stable and well understood drift chamber of this shear size. Gas purity, density and temperature all need to be carefully regulated and monitored to ensure charge gain is stable to the few percent level. For example, detailed non-linear corrections to the collected charge samples and rigorous quality cuts were needed in order to achieve the dE/dx resolution of $\sim$3% attained by the OPAL collaboration. Table 2.2 compares the parameters and performance of the LEP experiments.

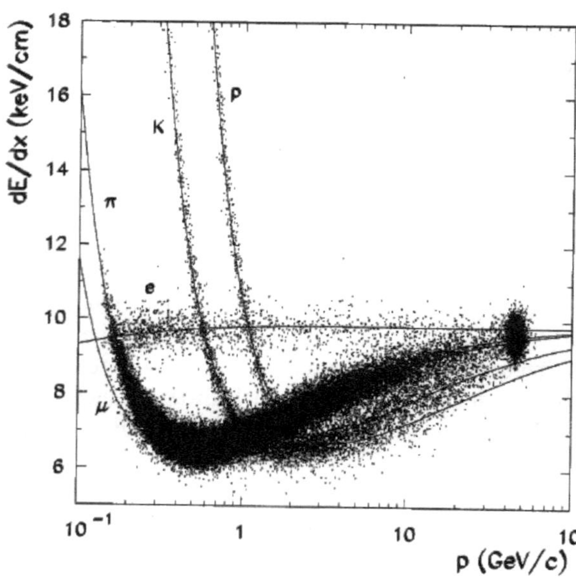

Fig. 2.7 Measurements from OPAL of the mean energy loss dE/dx of tracks in hadronic and di-muon decays of the Z^0. From [10]

Table 2.2 Comparison of dE/dx performance of the LEP experiments. Resolutions are quoted for single isolated tracks and in hadronic events (parenthesis) if available. After [9]

	ALEPH	DELPHI	OPAL
Reference	[11]	[12]	[10]
Detector type	TPC	TPC	Jet drift chamber
Size ($\phi \times L$)	3.6×4.4 m	2.4×2.7 m	3.6×4.0 m
Number of samples	338	192	159
Sampling length	4 mm	4 mm	10 mm
dE/dx resolution (%)	4.5	4.5	2.8

2.2.2 Čerenkov Ring Imaging

Worth a special mention are the Ring Imaging CHerenkov (RICH) detectors of the DELPHI experiment. The principle idea is to measure the velocity of a particle which is travelling faster than the velocity of light in the local medium (with refractive index n) by measuring the Čerenkov light cone opening angle (θ_C) and using the relationship $\theta_C = \frac{1}{n\beta}$. The velocity can then be combined with a measurement of the particle's momentum from the tracking detectors in order to determine the mass and hence identify the particle type. By the 1980s, the technique was reasonably well established and had been used in a collider environment by the UA2 experiment for electron ID. DELPHI[1] however proposed an altogether more sophisticated device consisting of a barrel and end-cap section each with a liquid (C_6F_{14}) and gas (C_4F_{10}, C_5F_{12}) radiator part able to separate e, μ, π, K and protons over a wide momentum range. Figure 2.8 illustrates the basic detection mechanism.

Figure 2.9 illustrates how the RICH detectors provide particle ID coverage from the relativistic rise region of the momentum range where separation of particle types by dE/dx becomes difficult. By using liquid radiator to cover the region up to a few GeV/c and gas radiator for the higher momenta, DELPHI were able to have a particle ID capability over almost the full momentum range of LEP 1. Software tags for particle types can be defined simply by making cuts on the dE/dx and RICH information depending on the purity and statistics of the sample required e.g. the efficiency for identifying $K^{\pm}$ averaged over the momentum spectrum above 0.7 GeV/c was about 70% with a contamination of about 30% while for protons the number was 70% efficiency for a contamination of 50%.

The stable operation and calibration of the DELPHI RICH was a huge technical challenge and many teething problems had to be overcome before the RICH data could contribute to physics results. A beautiful example of the power of a particle identification system containing a RICH capability is shown in Fig. 2.10. A candidate $B^- \rightarrow K^{0*}(892) + \pi^-$ decay is found in DELPHI by reconstructing the decay vertices in the silicon vertex detector and identifying the decay products with high probability by combining dE/dx and RICH information.

---

[1] Also, at about the same time, the CRID project of the SLD detector at SLAC.

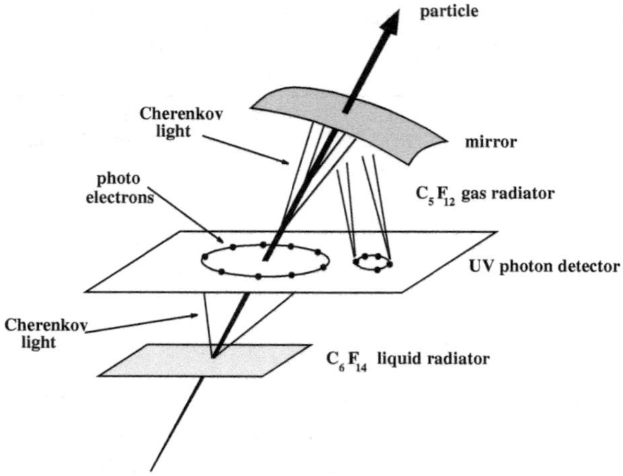

Fig. 2.8 The detection of Čerenkov light rings in the DELPHI RICH detectors, produced by the passage of charged particles through liquid and gaseous radiators

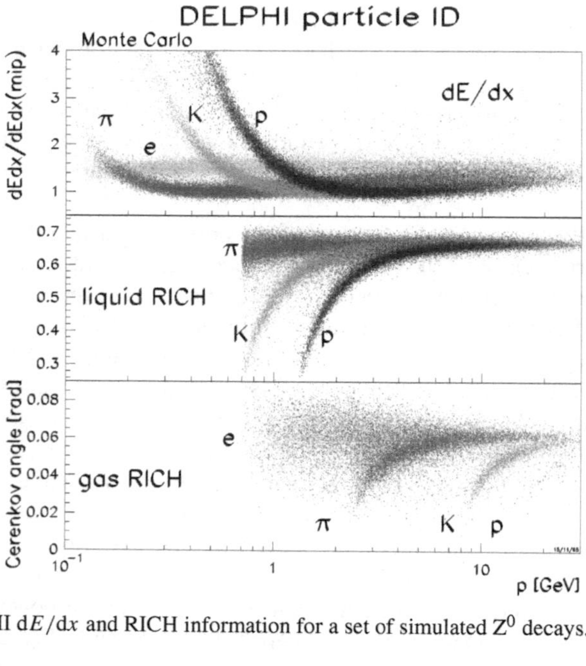

Fig. 2.9 DELPHI dE/dx and RICH information for a set of simulated Z^0 decays. From [12]

2.2.3 Overall Detector Performance

Although there was considerable overlap in the designs, there were also significant differences between the four experiments. These differences reflected the desire to cover all possible physics channels optimally and to get the correct balance between

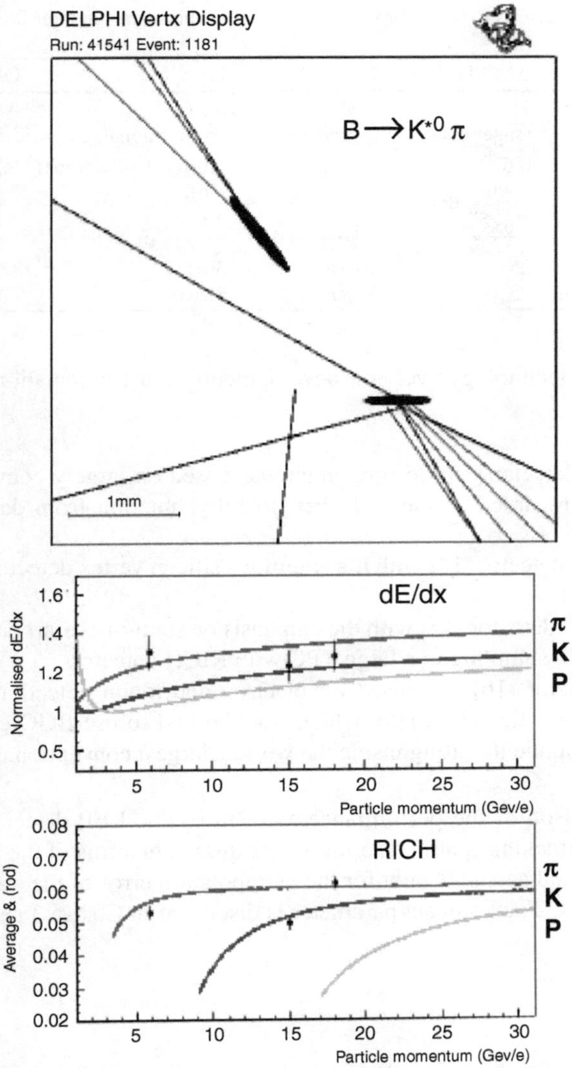

Fig. 2.10 A zoom-in view of the tracks extrapolated to the vertex region in the $x - y$ plane is shown in the *upper plot* where the Z^0 decay (*lower right*) and B decay (*upper left*) vertices are indicated by 3σ error ellipses. The plots below show the measured $\mathrm{d}E/\mathrm{d}x$ and RICH quantities for the decay products (π^- plus $K^+\pi^-$ from the decay of the $K^{0*}(892)$) overlaid on the curves expected for pions, kaons and protons. From [12]

Table 2.3 A comparison of the performance of the LEP experiments in their final configurations. After [17]

	ALEPH	DELPHI	L3	OPAL
B-field	1.5T (super cond.)	1.2T (super cond.)	0.5T (conventional)	0.435 (conventional)
$\sigma\,(1/p) \times \frac{10^{-3}}{\sqrt{\text{GeV}}}$	0.6	0.6	0.6(μ), 1.8(hadrons)	1.5
$\frac{\sigma(E)}{E}$(Emag.)	$\frac{0.18}{\sqrt{E\text{GeV}}} \oplus 0.01$	$\frac{0.32}{\sqrt{E\text{GeV}}} \oplus 0.04$	$\frac{0.02}{\sqrt{E\text{GeV}}} \oplus 0.01$	$\frac{0.06}{\sqrt{E\text{GeV}}} \oplus 0.02$
$\frac{\sigma(E)}{E}$(Hadronic)	$\frac{0.85}{\sqrt{E\text{GeV}}}$	$\frac{1.12}{\sqrt{E\text{GeV}}} \oplus 0.21$	$\frac{0.55}{\sqrt{E\text{GeV}}} \oplus 0.08$	$\sim \frac{1}{\sqrt{E\text{GeV}}}$
$R-\phi$ IP resolution	25 μm	20 μm	30 μm	18 μm
$R-z$ IP resolution	25 μm	30 μm	30 μm	20 μm

tried and tested technology over new developments in instrumentation. Examples of this include:

- The OPAL detector [13] design which was based on largely conventional technology and regarded as the 'safe bet' to fully function from day one of LEP operations.
- The ALEPH detector [14] with it's ambitious silicon vertex detector with double-sided readout.
- The DELPHI detector [15] with the emphasis on state-of-the-art particle ID combining $\mathrm{d}E/\mathrm{d}x$ sampling in a large TPC with RICH detectors.
- The L3 detector [16] optimised for electron and muon detection by including an electromagnetic calorimeter which was the first to use BGO crystals and all detector components sitting inside the world's largest conventional 0.5T magnet.

The comparison of the performance attained by the LEP detectors is given in Table 2.3. The tracking spatial resolutions are quoted in terms of the impact parameter resolution and accounts only for the extrapolation error of the track back to the production point. Track impact parameter is discussed in Chaps. 3 and 4.

References

1. A. Peisert: Silicon Microstrip Detectors. In: *Instrumentation in High Energy Physics*, F. Sauli (ed.). World Scientific, Singapore (1992)
2. J. Kemmer: Nucl. Instr. Meth. A **169**, 499 (1980)
3. G. Anzivino et al.: Nucl. Instr. Meth. A **243**, 153 (1986)
4. G. Batignani et al.: Nucl. Phys. B (Proc. Suppl.) **23**, 291 (1991)
5. V. Chabuad et al.: Nucl. Instr. Meth. A **368**, 314 (1996)
6. L3 SMD Collab., M. Acciarri et al.: Nucl. Instr. Meth. A **360**, 103 (1995)
7. P.P. Allport et al.: Nucl. Instr. Meth. A **346**, 476 (1994)
8. P. Chochula et al.: Nucl. Instr. Meth. A **412**, 304 (1998)
9. M. Hauschild et al.: Nucl. Instr. Meth. A **379**, 436 (1996)
10. M. Hauschild et al.: Nucl. Instr. Meth. A **314**, 74 (1992)
11. D. Buskulic et al.: Nucl. Instr. Meth. A **360**, 481 (1995)
12. DELPHI Collab., P. Abreu et al.: Nucl. Instr. Meth. A **378**, 57 (1996)

13. OPAL Collab., K. Ahmet et al.: Nucl. Instr. Meth. A **305**, 275 (1991)
14. ALEPH Collab., D. Decamp et al.: Nucl. Instr. Meth. A **294**, 121 (1990)
15. DELPHI Collab., P. Abreu et al.: Nucl. Instr. Meth. A **303**, 233 (1991)
16. L3 Collab., B. Adeva et al.: Nucl. Instr. Meth. A **289**, 35 (1990)
17. K. Mönig: Rept. Prog. Phys. **61**, 999 (1998)

Chapter 3
Experience in Reconstructing $Z^0 \rightarrow b\bar{b}$ Events

A tremendous amount of experience was gained as a result of extracting the most from the LEP data sets. In this chapter we review the developments that had the largest impact on b-physics analysis.

3.1 Impact Parameter and Decay Length

Central to studies of b-physics at LEP were two reconstructed parameters that were sensitive to the relatively long lifetime of b-hadrons: *impact parameter* and *decay length*.

Both parameters are schematically illustrated in Fig. 3.1 which shows a B-hadron produced at the primary vertex and flying a few millimetres before decaying into three particles at a secondary vertex. The impact parameter is the distance between the primary vertex and the track position at the point of closest approach and can be constructed as a 3D quantity or separately in the $R-\phi$ and $R-z$ planes. An important addition to this definition is the 'lifetime sign' . A positive (negative) sign is given to the impact parameter according to whether the point at which the track crosses the estimated b-hadron flight path, is upstream (downstream) of the primary vertex point[1]. With this definition, tracks from decays with significant lifetimes will have positive signs whereas particles originating from the primary vertex will have their tracks 'smeared' around the vertex point within the track reconstruction uncertainty and are equally likely to be signed positive or negative. Signing in this way is particularly useful since the negatively signed parameter distribution measures directly the parameter uncertainty. This follows since, apart from gross reconstruction errors, only tracks smeared by the detector resolution will contribute to the negative distribution. Decay products from b-hadrons had impact parameters of a few hundred microns in the LEP experiments and were typically measured with a precision of 20–50 μm depending on the track's momentum.

---

[1] This is strictly true for impact parameters in the $R-\phi$ plane. In 3-D the point of closest approach of the track to the b-hadron direction vector is taken since two tracks never cross in 3-D unless they collide with each other

G.J. Barker, *b-Quark Physics with the LEP Collider*, STMP 236, 37–55,
DOI 10.1007/978-3-642-05279-8_3,

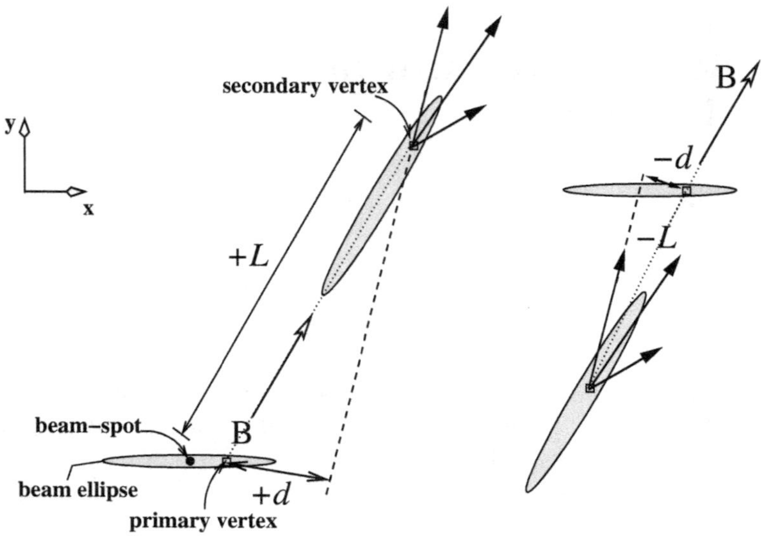

Fig. 3.1 The definition of lifetime-signed track impact parameter d and decay length L. For clarity, only the projection in the $R - \phi$ plane shown

The b-hadron decay length L is the distance from the event primary vertex to the b-hadron decay or secondary vertex point. In a similar way to the impact parameter, it is given a positive (negative) lifetime sign if the vertex is reconstructed in front (behind) of the primary vertex position with respect to the b-hadron direction. The LEP experiments were able to reconstruct decay lengths with an average resolution of ~ 300 μm dependent on the topology of the b-hadron decay.

In the early LEP years, effort was focused on reducing as far as possible the uncertainties in reconstructed impact parameter and decay length and later in understanding the details of the resolutions and accurately modelling them in simulation. In general, the uncertainty on the impact parameter has two contributions: tracking resolution and multiple scattering of the particle as it traverses the material of the detector. The multiple scattering component increases as the momentum of the particle decreases. Since b-hadrons decay well within the beam pipe of LEP, tracking was dominated by the most precise element of the tracking system, namely the silicon vertex detectors. Decay length uncertainty depends also on the intrinsic detector resolution and the effects of multiple scattering, but it also depends on the opening angle of the decay particles. A small opening angle will correspond to a large uncertainty in the crossing point of the tracks in the direction of the vector joining the primary and secondary vertices. A consequence of this is that the decay length uncertainty was observed to *increase* with momentum even though the multiple scattering component was disappearing due to the tendency of decay track opening angles to decrease as the momentum of the parent increases. This is illustrated in Fig. 3.2 which shows the uncertainty in decay vertex position along the flight direction for

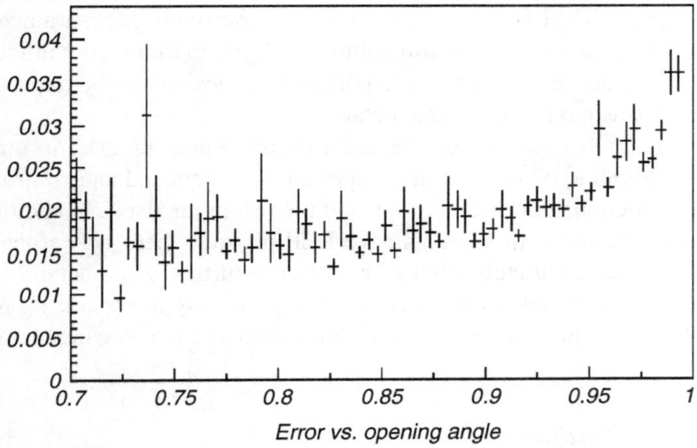

Fig. 3.2 The uncertainty in vertex position along the flight direction (in cm) for the decay $D^0 \to K\pi$ as a function of the cosine of the opening angle between K and π. Taken from [1]

a sample of $D^0 \to K\pi$ decays as a function of the opening angle between the K and π.

The impact of decay length and impact parameter resolution on physics measurements depends on the analysis details. For example, measuring particle lifetimes by decay length reconstruction would be based on the following relationship

$$t = m\frac{L}{p} \tag{3.1}$$

where t is the (proper) lifetime of a particle with momentum p and rest mass m. The uncertainty on the lifetime therefore depends on both decay length and momentum resolution. A fully reconstructed, exclusive, decay mode would in principle give the best momentum resolution but these analyses were rare at LEP because of the relatively low rate of b-production. A characteristic momentum uncertainty from inclusive reconstruction of b-hadrons was 15–20%. This improves for semi-leptonic decays (where scaling the measured momentum of the decay products can partially compensate for the loss of the neutrino) and in techniques using more sophisticated inclusive methods (discussed further in Sect. 7.2.4). The decay length uncertainty also contributes indirectly to the overall uncertainty. Lifetime analyses usually cut on the decay length to reduce backgrounds mainly from light-quark events. This introduces systematic uncertainties if the form of the L-distribution is not exactly known since the t-distribution of any background remaining in the sample after the cut must be accounted for in the final lifetime fit. In the DELPHI analysis of B^+ and B_d^0 lifetimes based on exclusively reconstructed semi-leptonic decays [2], most of the systematic error was due to uncertainties in the background estimate. A precise understanding of the resolution is even more important for analyses based on impact parameters mainly because badly measured tracks often result in large impact parameters. The LEP measurements of R_b based on impact parameters found that the quantities taken from the simulation e.g. light quark tagging efficiency and hemisphere correlations soon dominated the systematics unless the impact parameter uncertainty was modelled in great detail.

In summary, as b-physics measurements became more precise, in order to go further it was found to be increasingly important to understand and control the tails of resolution functions rather than the size of the resolution itself. In addition, given the importance of simulation to inclusive LEP physics, there was a need to also model these effects accurately. Below, we discuss further some details of hit and track reconstruction which were addressed along the way to optimising resolution and we outline some methods of tuning simulation to be a realistic match to the data.

3.1.1 Cluster Finding

If the charge released by the crossing of a charged particle through a silicon-strip device is collected on only one strip, the resolution on the crossing point (σ) will be determined by the strip pitch r: $\sigma \sim r/\sqrt{12}$ (which follows from the standard deviation of a uniform distribution). Better resolution can be achieved if the charge is spread between a few strips and some kind of mean position for the charge is constructed by weighting the strip positions. For example, if the charge is shared between two strips located left and right of the crossing point, the position of the crossing point x from some nominal origin is given by,

$$x = r(i_L + \eta) \tag{3.2}$$

where i_L is the left strip position and

$$\eta = \frac{PH_R}{PH_L + PH_R} \tag{3.3}$$

where PH_L, PH_R are the pulse heights of the left and right strips respectively. The resolution on the crossing point for this situation goes as: $\sigma \sim r/SN_{cl}$ where SN_{cl} is the sum of the signal-to-noise ratio of the two strips in the cluster which is typically 10–30. Specialised clustering algorithms were developed by the LEP experiments to optimise position resolution for specific devices and geometries (see e.g. [3, 4]). Worth a mention here is a detailed study of clustering by DELPHI [5] which concluded that the linear charge sharing model of (3.3), which assumes a rectangular pulse shape, could be improved on by expecting a flat-top distribution plus rectangular sides. Improved resolution was particularly marked in the $R - z$ plane where the charge could be shared across many strips if the particle crossed at small polar angles.

A general problem that all clustering algorithms had to address was that of *cluster splitting* where the underlying charge cluster is reconstructed as two or more clusters. Reasons for this include the presence of dead strips, excessively noisy strips that are not read-out by the data acquisition algorithm or strips whose S/N ratio drops below the threshold for readout because of a small crossing angle spreading the charge over many strips. In the case of DELPHI these problems were addressed by first setting up a detailed database of dead strips and e.g. merging together clusters that are found to be separated by dead strips (for details see [6]). Although such cluster merging only affected a few percent of all clusters it was important in reducing the effects of non-Gaussian tails in resolution distributions. This same database would then also be applied to the simulation so that inefficient regions of the detector were accurately modelled.

3.1.2 Track Fitting and Pattern Recognition

Developments in this area mainly centred on how best to combine hit information across sub-detectors that varied greatly in their size and precision. Much of this work was specific to a particular collaboration but we highlight in this section two aspects of tracking where independent investigations made by the collaborations prompted the same action.

The first concerns the track fit itself. Pre-LEP, such track fits were typically made using a simple least-squares minimisation fit to the hit set and early versions of the LEP tracking algorithms worked the same way. These algorithms were found to work well for isolated tracks but were badly degraded in the rather dense track environment of LEP jets or multi-track tau decays, where the chance of the pattern recognition associating wrong hits increases. These wrong hits or 'outliers', so-called because they often appeared rather far from the true trajectory, can significantly bias the result of a simple least-squares fit that treats them with equal weight to the correctly assigned hits. The LEP collaborations found that a more robust approach was given by Kalman filter algorithms which work recursively extrapolating the track vector successively through the hit set, each time updating by taking the weighted mean with the measured hit (see e.g. [7]). Care must be taken in the extrapolation to correctly account for the effects of multiple scattering and energy loss $\mathrm{d}E/\mathrm{d}x$. This type of algorithm subsequently became the default tracking approach and was also used widely for decay vertex fitting applications (e.g. [8]).

The second aspect of tracking which saw general development concerns pattern recognition. Tracking was dominated by conventional gaseous drift chambers in the early years of LEP. This was due to the fact that it took time for ALEPH and DELPHI to fully understand and integrate the data from their new silicon vertex detectors into physics analysis and L3 and OPAL began data taking without silicon devices. Once the silicon devices were properly integrated and, in particular, accurately aligned (see Sect. 3.1.3 below), it became clear that the strategies used up to that point to form tracks did not use optimally the superb precision of the silicon detectors. The reason was that most pattern recognition algorithms worked from the 'outside-in' i.e. track elements were first found in the surrounding gaseous tracking chambers and extrapolated in towards the silicon vertex detectors. Depending on the circumstances, the extrapolation errors from this procedure could be as large as a few mm's, making the possibility of associating the wrong (or no) silicon hits to the track rather high. The effects were particularly noticeable in jets where particles are closely bunched together. Algorithms were therefore changed so that the most precise information available, i.e. the hits from the silicon-strip devices, became an integral part of the pattern recognition step of track reconstruction. As an example, we consider the changes made to the DELPHI tracking algorithm. In addition to the existing Outside-In track searches, DELPHI added searches that started from hit sets in the silicon and tried to match them to track elements in the surrounding tracking detectors: the Inner Detector/TPC/Outer Detector with intrinsic resolutions of $\sim 60/250/300\ \mu$m respectively. The Kalman track fitter was run on all combinations of associated hits and those with a poor fit quality rejected. Any remaining ambiguous solutions were removed by demanding that the 'correct' solution be that containing the most silicon hits or composed of the most track elements. A illustration of the effect that these changes had on event reconstruction is shown in Fig. 3.3 where hits in the silicon detectors are depicted as squares, track fits are the solid lines and the extrapolation of the tracks towards the interaction point are dotted lines. The old (left plot) and new (right plots) algorithms have processed the same data event which is a rather typical hadronic Z^0 decay producing two or more dense jets of particles. The result is that the new algorithm finds two extra tracks in the bottom hemisphere jet and three extra in the upper hemisphere. In addition, what appear to be false tracks of high curvature not pointing back to the interaction point that appear in the first picture are absent in the new results.

On average, DELPHI found that the new algorithm gave about four extra tracks per event that contained silicon hits and had acceptable fit-χ^2 values and the yields (and resolutions) of all mass peaks improved greatly e.g. (see Fig. 3.4). The result

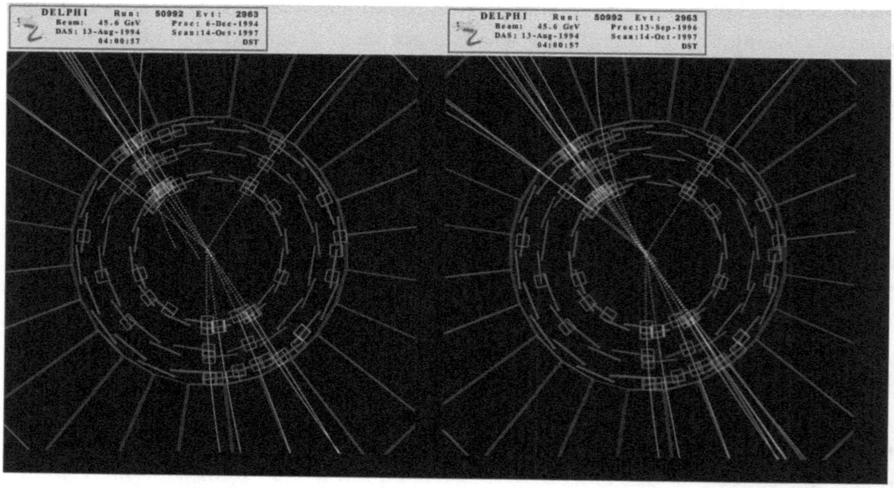

Fig. 3.3 A direct comparison of running the old (*left*) and new (*right*) track reconstruction algorithm of DELPHI on the same data event

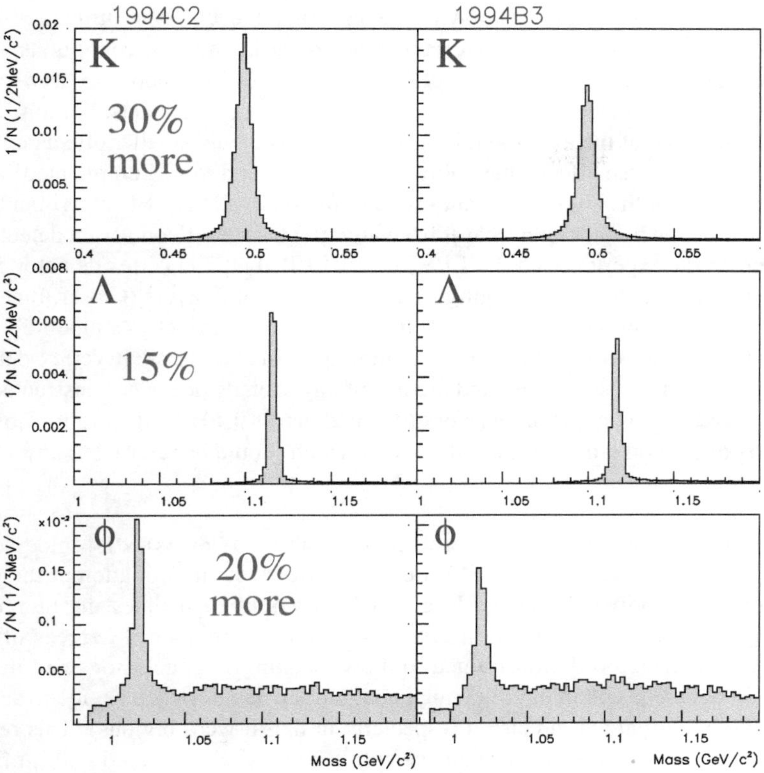

Fig. 3.4 Comparing the effect of the old (*right column*) and new (*left column*) DELPHI tracking on some reconstructed mass peaks. The numbers quoted refer to the increase in yield seen

of similar changes made to the tracking algorithms of the ALEPH collaboration resulted in a reduction of pattern recognition errors in 3-prong tau decays by more than 50% from the previous value of approximately 30%. They also reported large gains in the mass and vertex resolution of well-reconstructed decays.

3.1.3 Vertex Detector Alignment

The subject of silicon vertex detector alignment, i.e. knowledge of where in space individual silicon strip devices are relative to each other and to the surrounding experiment, is vital if the sub-10 μm intrinsic resolution of such devices is not going to be severely degraded.

The alignment process usually starts with a 3D survey based on an optical and/or mechanical system measuring the positions of reference features built into the structure of each detector element. Such surveys can be precise to a few x10 μm. Once in position, knowledge of where the complete vertex detector is with respect to the rest of the experiment may only be known at the millimetre level. To go further, typically one must make use of the fact that reconstructed tracks must be continuous across the boundaries of neighbouring devices and layers of devices and also between the vertex detector as a whole and it's surrounding tracking components. For any detector as knowledge of the alignment improves, so smaller and smaller effects must be understood and these effects invariably become more and more complicated!

Since many of the alignment issues were common to all the LEP experiments, we will illustrate the main lessons learnt from aligning a large silicon vertex detector by focusing on the experiences of DELPHI. The DELPHI silicon vertex detector, introduced in Sect. 2.1, in its final configuration consisted of 3 layers (Closer, Inner and Outer) each containing 24 overlapping modules (4 strip devices to a module). After more than 5 years of development, the alignment of the DELPHI vertex detector was arguably the most detailed and precise of any such device ever constructed.

An interesting outcome of this work (detailed in a DELPHI collaboration note [9]) was a set of general guidelines for alignment which should be relevant to any similar application:

(1) The analysis should solve the problem of aligning 3 (layers) $\times$ 24 (modules) $\times$ 6 (3 positions, $\times$2 angles) $=$ 432 degrees of freedom and not attempt the problem at the individual device level – which would be 4-times the number of degrees of freedom. Clearly a system with a huge number of degrees of freedom becomes too ill-constrained to draw meaningful conclusions and there is the practical problem of collecting enough track data for each device to achieve sufficient alignment precision (especially in the forward regions). This recommendation assumes that each device within a module is surveyed with sufficient precision and that the mechanical design is rigid enough to guarantee stability over the lifetime of the experiment.

(2) The internal alignment of the vertex detector should be completed before attempting to align the resulting rigid body externally with respect to the rest of the experiment. This condition is a result of the fact that tracking volumes surrounding the vertex detector would typically rely on the precise vertex detector space points as part of their own calibration. Projecting any external misalignments into the internal alignment of the vertex detector would be difficult to unfold at a later stage.

(3) A design that includes an overlap region between adjacent modules allows the reconstruction of track residuals which are invaluable for alignment studies (see below).

(4) Any algorithm attempting to simultaneously estimate 432 parameters is doomed to failure since (a) it is difficult to find a global minimum in such a multi-parameter space and (b) it is not reasonable to assume that all local (and global) distortions to the ideal geometrical configuration could be taken account of. An iterative approach was therefore adopted whereby the Outer layer was aligned first and acts as the *master* layer so that individual modules in the Inner and Closer layers could then be aligned with respect to the corresponding Outer modules. Approaching the job in this order ensures that the information from

the inner layers, which suffers less from multiple scattering and is hence more precise, is present to ensure that the track extrapolation is as reliable as possible.

(5) Care should be taken to identify and unfold any distortions at the level of the complete vertex detector. Because no structure is infinitely mechanically rigid, these movements will always occur at some level and are a particular worry during transport and installation into the experiment. In DELPHI's case, the detector was split longitudinally into two half-shells and the alignment studies revealed effects consistent with the two halves, when inside DELPHI, being skewed relative to each other as a function of z so that their edges were no longer parallel. In addition each half had undergone a 'twist' deformation along the z-axis similar to the effect of rotating each end-cap simultaneously in the opposite sense. Some parameters, such as the z position of individual modules were found to have moved by up to ten times the precision of the initial survey.

The DELPHI approach was based on 3 types of track from Z^0 decays: hadronic tracks passing through the overlap region of two adjacent sectors ('overlap tracks'), hadronic tracks contained within one sector ('3-hit tracks') and back-to-back muons from Z^0 decay passing into oppositely facing sectors ('$\mu\mu$ tracks'). Note that muon-pair tracks are ideal for alignment purposes since their high momentum ensures they are free from the effects of multiple scattering and they are isolated making them essentially immune from pattern recognition errors common in dense track environments. The addition of (high momentum) hadronic tracks is however essential in order to boost the statistics of the alignment track sample.

From these tracks 3 types of residuals were defined (see Fig. 3.5) and it was the goal of the procedure to *minimise the spread* of the residuals by adjusting the correlated parameters. Outer layer alignment in the $R-\phi$ plane was based on overlap tracks where a circle was constrained to pass through a hit in the Closer or Inner layer and either of the two hits in the Outer layer overlap region. The residual is then the distance between the unused hit in the overlap region and the position predicted for this hit by the track fit. The corresponding quantity in the $R-z$ plane substituted $R-\phi$ hits with $R-\phi+R-z$ hit doublets and formed residuals to a helix trajectory. The residual to 3-hit tracks was defined by passing a circle through hits in the Closer and Outer layers and forming the residual with the Inner layer hit. For $\mu\mu$ tracks a circle (or helix) was constrained to pass through 2 $R-\phi$ hits (or hit doublets) in opposite sectors of the Outer layer and the residual was formed to the $R-\phi$ (or hit doublet) in either of the 2 corresponding opposite sectors of the Closer layer. Note that the global alignment of the vertex detector within the rest of DELPHI could be fixed by muon-pair tracks by demanding that the tracks are continuous across the interfaces between the various tracking systems.

Once the vertex detector alignment precision approaches the intrinsic tracking precision of the detector itself, secondary effects become important and must be handled if the alignment precision is to improve further. We consider now three such examples from the DELPHI analysis: Lorentz shift, barycentric shift and module bowing.

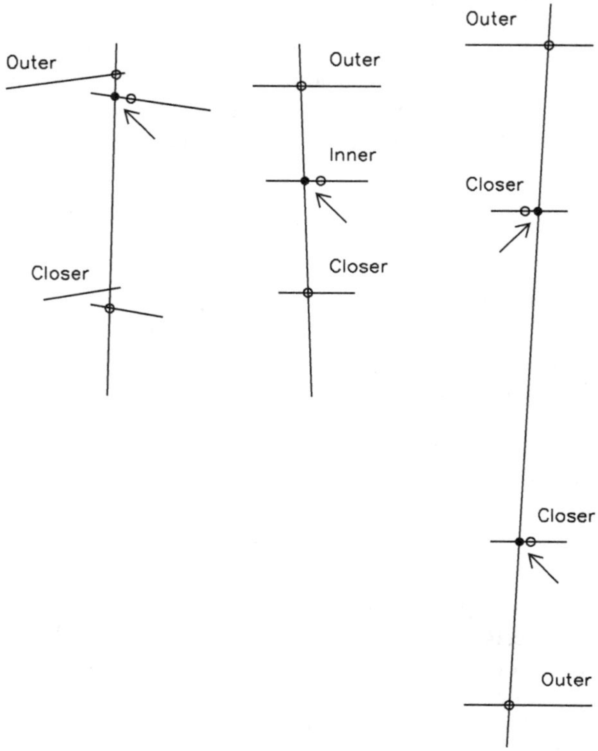

Fig. 3.5 Schematic representation of residuals in the $R-\phi$ plane for overlap, 3-hit and $\mu\mu$ tracks. *Open circles* represent recorded hit positions and *black dots* represent hit positions predicted by a fit through the recorded hits (the small curvature of tracks is not shown)

The DELPHI solenoidal magnetic field aligned with the z-axis caused holes and electrons to drift at a Lorentz angle to the electric field direction inside the strip devices. The result was a shift in the reconstructed $R-\phi$ coordinates away from the true crossing point of the charged particle (the $R-z$ coordinates are unaffected) which would propagate through to an alignment uncertainty unless accounted for. Since the Lorentz shift has the same effect on all devices, it would have been impossible to unfold without the DELPHI flipped module design introduced in Fig. 2.5. The direction of the Lorentz shift flips if the direction of the electric field is reversed, as happens between adjacent devices in a flipped module. This means that e.g. a track that passes through a device in the Closer layer and a device in the Outer layer which has been flipped relative to the Closer device, will form a residual at the Inner layer which is sensitive to the size of the Lorentz shift - tracks passing through devices with electric fields all in the same direction will form residuals that are completely unaffected. The top plot of Fig. 3.6 shows the $R-\phi$ 3-hit residual as a function of the z coordinate where two jumps appear exactly at the boundaries

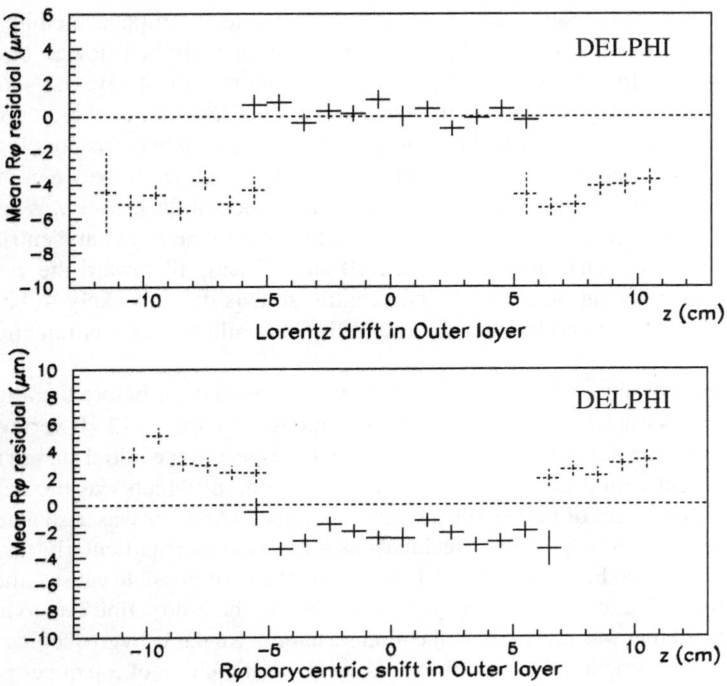

Fig. 3.6 (*top*) Shows the mean residual in $R - \phi$ for 3-hit tracks as a function of z at the Outer layer. The Lorentz drift effect is evident at $|z| = 6\,\text{cm}$ where the flipped module design changes the E-field direction. (*bottom*) Shows mean residuals in $R - \phi$ for overlap tracks as a function of $z$at the Outer layer. Barycentric shift induces jumps at the flipped boundaries

between flipped devices in a module and which corresponds to a Lorentz shift of around 6 μm.

Barycentric shift is a term coined by DELPHI to describe the effect of ionisation charge not being uniformly released along the trajectory as a charged particle traverses the silicon bulk material. As a consequence, the centre of gravity of the charge cloud arriving at the readout strips no longer corresponds to the mid-plane of the device resulting in a systematic shift in the reconstructed particle crossing coordinates. For the effect to be visible the tracks must have a crossing angle away from normal incidence and it increases in size as this angle becomes more extreme. In the $R - \phi$ plane adjacent modules in a layer were inclined at a relative angle of 15° so that the barycentric shift is different in each and therefore potentially visible in the hit residuals of tracks traversing the overlap region between modules. It is important to realise that this effect could be absorbed in a radial shift of modules closer or further away from the interaction point so that the device mid-plane is forced to coincide with the centre of the charge distribution. If however the radial positions are already precisely known from a survey, as was the case with DELPHI, the freedom to do this does not exist. The size of the effect is relatively small and was only made possible in DELPHI due to the flipped module design which meant that the sign of the shift flipped at each flipped device boundary, effectively doubling the size of the effect. The bottom plot of Fig. 3.6 shows the mean $R - \phi$ overlap residual in the Outer layer as a function of the z-coordinate of the track along that layer and two distinct jumps are visible corresponding to the boundaries between flipped devices. Given a relative inclination between modules of 15° a barycentric shift of 10.5 μm towards the p-side of the device was deduced. Using such techniques, DELPHI were eventually able to measure barycentric shifts of holes and electrons ranging between 10 and 20 μm, all toward the p-side of devices. The physical origin of the barycentric shift is thought likely to be due to variations in dopant concentration through the bulk silicon but this remains to be confirmed.

Both the Lorentz and barycentric shifts were corrected for before the alignment procedure was run i.e. before the positions of modules in the DELPHI vertex detector database were altered to reflect the shifts with respect to the initial survey results. The alignment results contained some surprises - one of which was the degree of global misalignment of half shells already mentioned. Another was associated with a U-shape of the $R - \phi$ overlap residuals as a function of z particularly for Closer layer modules (see Fig. 3.7). After a long investigation of possible causes, the effect was finally traced to a bowing of the modules along the z-direction due to a kevlar stiffening rail that was glued along the module length. Kevlar is hygroscopic and was resulting in a bi-metal effect which would increase at the rate of 3 μm per percent-change in humidity. Figure 3.7 shows how the effect was successfully removed by correcting the bowed shape as a function of humidity.

This section has illustrated the incredible level of detail to which one must consider the alignment of vertex detectors if the superb intrinsic precision of these

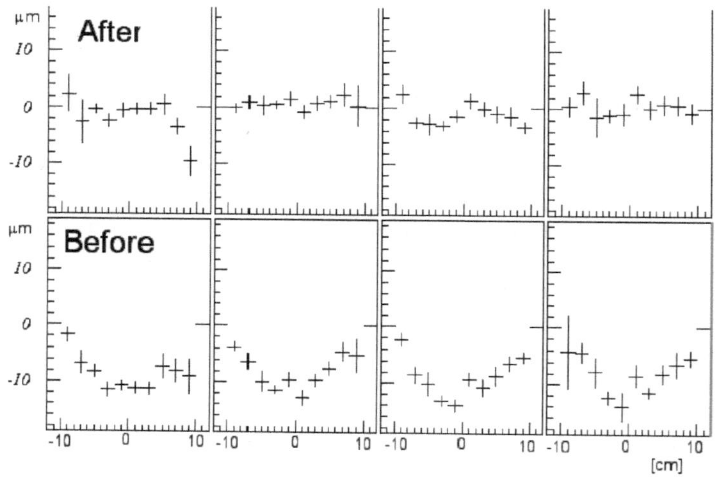

Fig. 3.7 Examples of mean $R - \phi$ residuals of overlap tracks in four modules as a function of z both before and after a correction for bowing has been applied

devices is not to be ‘smeared out’ following track reconstruction. It is probably fair to say that the level of detail to which the DELPHI vertex detector alignment was eventually understood was unprecedented and there were many useful lessons learnt along the way which should find a wider application than just LEP data analysis. Although difficult to quantify, DELPHI estimated (from methods such as the miss-distance of muon-pair events at the primary vertex) that the alignment precision achieved was of the order of half the intrinsic precision i.e. around 4 μm in $R - \phi$ and 4 μm in z (measured at $\theta = 90°$). Furthermore, the surrounding tracking volumes of DELPHI were able to benefit their own calibration by using tracks extrapolated from the vertex detector as reference. For instance, it was possible to correct for many systematic effects in the DELPHI TPC in this way.

3.1.4 Single Hit Uncertainties

As discussed in Sect. 3.1, precision measurements based on impact parameter and decay length were eventually limited by a lack of control of non-Gaussian tails in resolution functions. In addition to alignment uncertainties which were discussed above, a main contributor to these effects was the assignment of uncertainties to hit positions in the silicon-strip devices which entered the track fits. To fully account for all effects was an enormous undertaking and the level of detail to which hit uncertainties were parameterised varied greatly from collaboration to collaboration – driven by the level of precision demanded by measurements such as R_b. In the case of DELPHI, hit uncertainties were parameterised as a function of crossing angle, cluster pulse height and number of contributing strips, strip noise and the charge sharing, η. As an example, Fig. 3.8 shows how the hit resolution as measured by track residuals varies strongly as a function of three categories of cluster defined by pulse height and the number of strips in the cluster.

3.1.5 Resolution Tuning

Although the use of simulation is kept to a minimum, inclusive b-physics analyses almost always involve the comparison of data to simulation at some point to estimate a parameter inaccessible directly from the data. Clearly the level of agreement between data and simulation has to be good to avoid large systematic uncertainties in parameters estimated this way. The detail with which vertex detectors were modelled for LEP analyses tended to increase year-on-year and the number of effects included would increase as knowledge and understanding of the data grew. The final, large, simulation data sets of the LEP collaborations would have contained e.g. a detailed description of the vertex detector material (in some cases down to the last screw), a map of all dead, inefficient and noisy regions, Lorentz shifts in ionisation charge collection and the application of random shifts in module positions

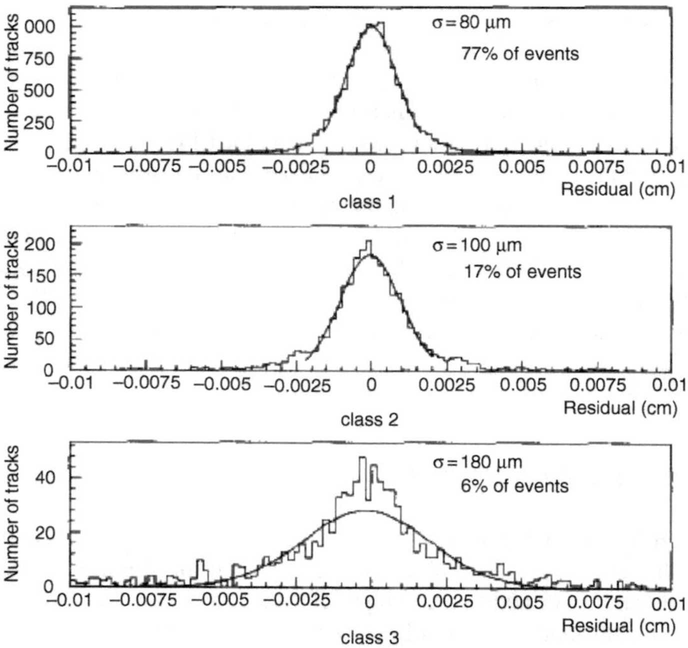

Fig. 3.8 For three types of cluster/hit in the Inner layer of the DELPHI Vertex Detector, the plots show the residual of this hit to a track passing through hits in the Closer and Outer layers. To convert residual to a hit precision multiply by 0.82. Taken from [3]

in order to simulate the effect of residual misalignments in the data. The simulated data would then be fed through *exactly* the same reconstruction code as the data. Despite this, disagreement between data and simulation at some level is unavoidable and tuning of the simulation was essential in order to account for any residual mismatches.

As part of the effort to measure R_b with ever greater precision, the ALEPH collaboration found it necessary to tune both the impact parameter resolution and reconstruction efficiency of tracks containing vertex detector hits. An observed access of such tracks in simulation was randomly stripped away as a function of track momentum, polar angle θ and the number of vertex detector hits in the $R-\phi$ and $R-z$ planes. The impact parameter resolution was *smeared* by making shifts sampled at random from exponential functions (Gaussian smearing was not found to give good agreement with the data). The parameters of the smearing were also determined as a function of momentum, θ and the number of vertex detector hits in the $R-\phi$ and $R-z$ planes. Interestingly, it was also found necessary in 0.5% of hemispheres, to smear all tracks together by the same amount. This was attributed to cases where the primary vertex was incorrectly reconstructed which would have a common effect on all tracks in the event. Details of the ALEPH procedure are given in [10] and an illustration of its effectiveness is shown in Fig. 3.9 where the

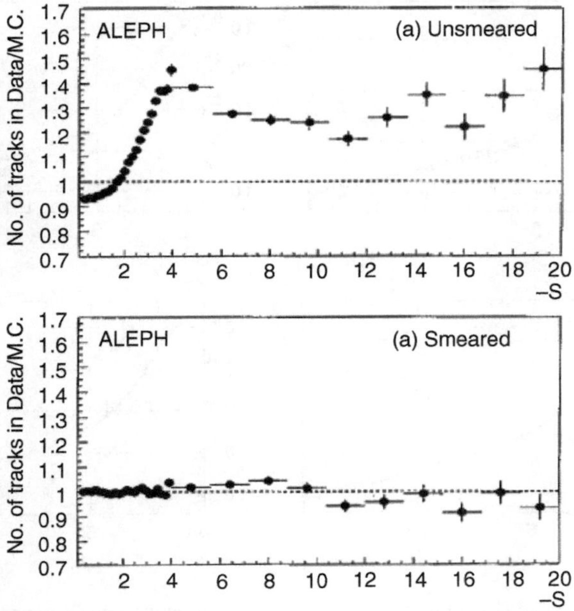

Fig. 3.9 The ratio of the number of tracks observed in data relative to simulation as a function of their negative impact parameter significance (**a**) before and (**b**) after applying the simulation tuning. From [10]

track multiplicity as a function of impact parameter significance ($-S$) [2] becomes well modelled in simulation after applying the tuning.

In DELPHI, a detailed procedure to tune the resolution in $R - \phi$ was developed and is detailed in [11]. The subsequent upgrade of the vertex detector to 2D readout resulted in the extension of the procedure to the $R - z$ plane. It was found that the $R - \phi$ and $R - z$ projections could be treated essentially independently of one another (see [12]) in the following procedure:

- The negative lifetime-signed impact parameter distributions ($\delta_{R-\phi}, \delta_{R-z}$) are parameterised.
- The numerical coefficients of these parameterisations are determined through fits to the data.
- The uncertainties of $\delta_{R-\phi}$ and δ_{R-z} are corrected in simulation and data according to the parameterisation but the correlation between $\delta_{R-\phi}$ and δ_{R-z} is left unchanged.
- Small additional smearing is applied to the impact parameters in $\delta_{R-\phi}$ and δ_{R-z} in simulation to achieve the best possible match to the data.

[2] Impact parameter significance refers to the impact parameter scaled by its uncertainty. Impact parameters with a negative lifetime sign are direct measures of resolution.

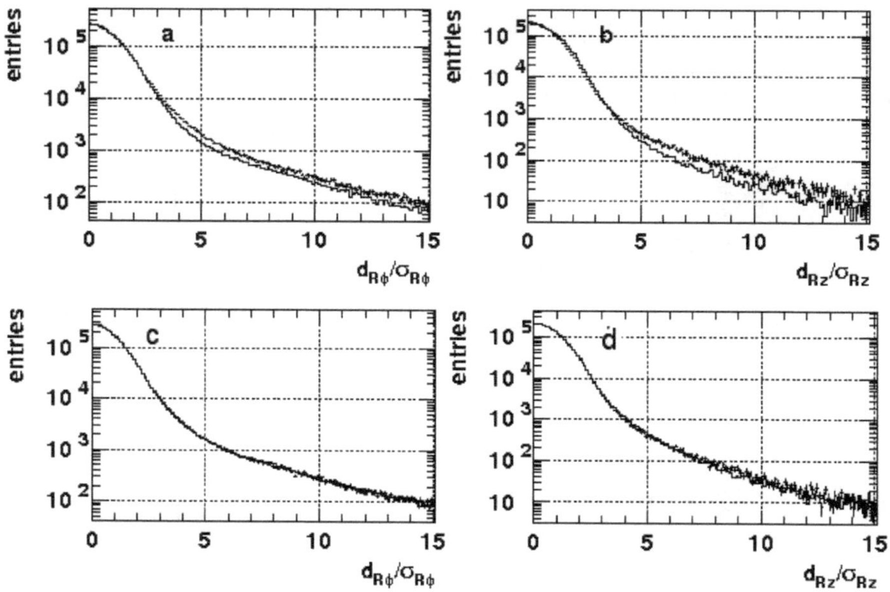

Fig. 3.10 Plots (**a**) and (**b**) compare $R - \phi$ and $R - z$ negative impact parameter significance distributions in data (*points*) and simulation (*histogram*) before the DELPHI resolution tuning procedure and (**c**) and (**d**) show the situation after tuning. From [13]

Figure 3.10 shows the improvement that this procedure made to precisely reproducing the impact parameter resolution in simulation. Good agreement can be seen between data and simulation even out to the extremes of the distribution tails.

3.2 Particle Identification

Particle identification (PID) played an important role in LEP b-physics. The presence of leptons with high transverse momentum were the key tag of b-quarks in the early publications and the identification of kaons and protons enabled tags of strange b-hadrons and baryons in the more sophisticated inclusive studies which followed. The LEP experiments had at their disposal a variety of measurement tools for PID including $\mathrm{d}E/\mathrm{d}x$ and Čerenkov Ring Imaging that were discussed in Sect. 2.2. Ultimately the best PID performance was obtained by the combination of complimentary measurements by methods such as neural networks (which we briefly review in Sect. 7.1.1). Below are a few representative examples of how these PID tags were built up and their performance.

3.2.1 Lepton Identification

Electrons are not easy to identify cleanly by cutting on a single quantity but there are a number of measurable discriminators in LEP-type detectors which when combined in an optimal way, can give a powerful electron signature. In addition to $\mathrm{d}E/\mathrm{d}x$ and RICH tags, there is extra information available from E/p (requiring a match between an external track of momentum p and a calorimeter cluster of total energy E) and electromagnetic shower shape since electron initiated showers tend to be rather narrow at the start of the longitudinal shower development.

Combining such quantities with neural network algorithms became the preferred analysis method for lepton ID among the LEP collaborations. OPAL for example developed an electron ID network based on twelve variables, variations of which were used in R_b [14] and $A_{\mathrm{FB}}^{\mathrm{b}}$ [15] analyses. Figure 3.11(a) from the OPAL R_b analysis shows the electron ID network output and the discriminating power against hadrons faking electrons and electrons from photon conversions. In fact to reduce the conversion background, OPAL produced a separate network trained on conversion electrons and required that there be a partner electron present. By requiring electron candidates to have a conversion network value less than 0.5, the identification efficiency for electron from b-hadron decay was about 68%.

In contrast to electrons, muons are experimentally straightforward to identify and do not require use of complicated combined tagging variables. Matching track segments from muon chambers with tracks reconstructed in the central tracking detectors coupled with a loose cut on $\mathrm{d}E/\mathrm{d}x$ to reduce kaon background, is

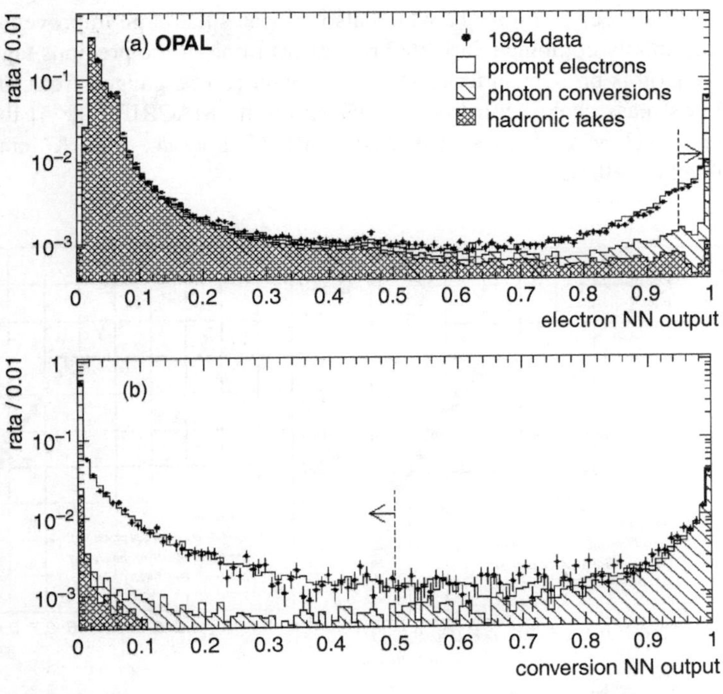

Fig. 3.11 Outputs of the OPAL (**a**) electron ID network and (**b**) photon conversion electron network. *Dashed lines* indicate cuts used in the analysis to define the final electron sample. From [14]

sufficient to reconstruct $b \to \mu$ with an $\sim$ 80% efficiency. The remaining background is usually from a hadron 'punching through' from the hadronic calorimeters and can be reliably estimated using simulations.

3.2.2 Hadron Identification

Kaon and proton identification is relatively difficult to address experimentally. The main tool used by the LEP experiments was dE/dx and in Fig. 2.7 we saw the level of discrimination that measurements from the large tracking chamber of the OPAL detector can produce. The DELPHI detector was at a distinct advantage for this type of analysis because of a Čerenkov Ring Imaging capability, discussed in Sect. 2.2.2, and we have seen (Fig. 2.9) how the RICH detector effectively extends the PID capability above momenta of around 1GeV/c. The DELPHI collaboration were able to take full advantage of this situation by building on algorithms already established for dE/dx and RICH with two main developments: (1) combining tag quantities that are weakly correlated with neural network techniques, and (2) complimenting the dE/dx information from the main tracker with that from the silicon vertex detector (which adds only a few points but with good energy resolution). The resulting kaon and proton tags, implemented in a package called MACRIB [16] found application in some of the final, high precision DELPHI inclusive analyses. The tagging performance can be seen in Fig. 3.12 which also indicates the large improvement that the more optimally combined variable brought over one of the previous tags, RIBMEAN. What this brought in terms of physics gain can be gauged from Fig. 3.13 which shows mass peak reconstructions based on the MACRIB tag. At the same background level, MACRIB reconstructed over 50% more $\phi \to K^+K^-$ and more than doubled the efficiency for $\Lambda \to p\pi$.

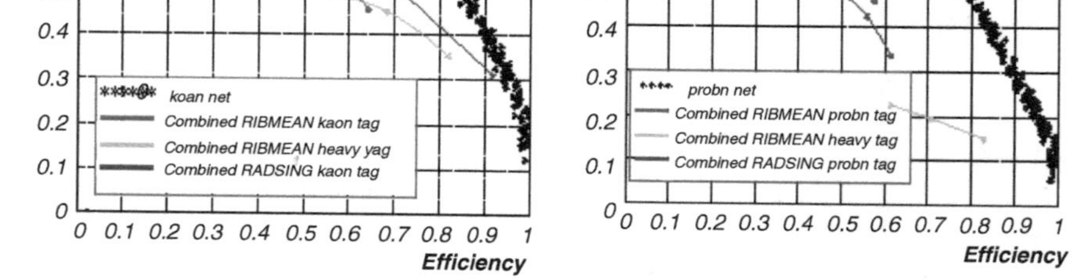

Fig. 3.12 Purity vs efficiency for correctly tagging (*left*) kaons and (*right*) protons with the MACRIB package. The improvement over the RIBMEAN tag is also indicated. From [16]

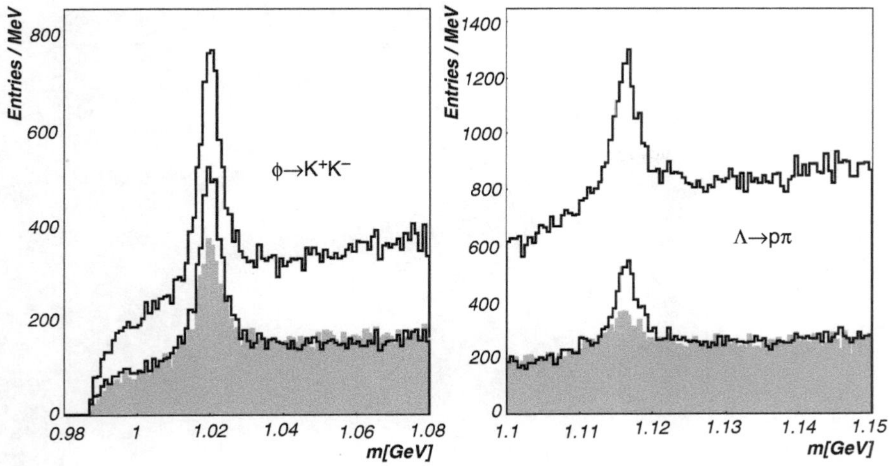

Fig. 3.13 The improvement in ϕ and Λ hadron reconstruction seen in DELPHI data by applying the MACRIB kaon and proton tags (*open histogram*) compared to RIBMEAN (*shaded histogram*). In each case, the comparison is made at the same level of background and at the same purity: (*left*) for $\phi \rightarrow \mathrm{K}^+K^-$ and (*right*) for $\Lambda \rightarrow \mathrm{p}\pi$. From [16]

References

1. A. Nomerotski: Advantages and Limitations of Microvertex detectors. In *Proceedings of the 4th International Workshop on Vertex Detectors*, 12–16 June 1995, Eingedi, dead Sea, Israel
2. DELPHI Collab., P. Abreu et al.: Z. Phys. C **68**, 13 (1995)
3. DELPHI Collab., N. Bingefors et al.: Nucl. Instr. Meth. A **328**, 447 (1993)
4. P.P. Allport et al.: Nucl. Instr. Meth. A **346**, 476 (1994)
5. P. Mattsson: Improvements to the Cluster Reconstruction Algorithm for the Vertex Detector. DELPHI Collab. Note 95-147 MVX 6 (1995)
6. A. Nomerotski, I. Stavitski and K. Österberg: Cluster Repairing in the Microvertex Detector. DELPHI Collab. Note 96-134 MVX 17 (1996)
7. R. Fr'uhworth: Nucl. Instr. Meth. A **262**, 444 (1987)
8. DELPHI Collab., P. Abreu et al.: Z. Phys. C **53**, 567 (1992)
9. V. Chabaud, A. Andreazza, P. Collins and H. Dijkstra: Alignment of the DELPHI Vertex Detector. DELPHI Collab. Note 95-177 MVX 10 (1995)
10. ALEPH Collab., R. Barate et al.: Phys. Lett. B **401**, 150 (1997)
11. G. Borisov, C. Mariotti: Nucl. Instr. Meth. A **372**, 181 (1996)
12. G. Borisov, C. Mariotti: Tuning of the Track Impact Parameter Resolution of the Upgraded DELPHI Detector. DELPHI Collab. Note 97-95 PHYS 717 (1997)
13. DELPHI Collab., J. Abdallah et al.: Eur. Phys. J. C **32**, 185 (2004)
14. OPAL Collab., G. Abbiendi et al.: Eur. Phys. J. C **8**, 217 (1998)
15. OPAL Collab., G. Alexander et al.: Z. Phys. C **70**, 357–369 (1996)
16. Z. Albrecht, M. Feindt and M. Moch: MACRIB- High Efficiency, High Purity Hadron Identification for DELPHI. hep-ex/01111081 (1999)

Chapter 4
Tagging $Z^0 \rightarrow b\bar{b}$ Events

Reconstructing from the data a quantity (the 'tag') which is correlated in some way to the presence of a b-quark in the underlying event, played a part in almost all b-physics analyses at LEP. Such tags were used in different ways:

- b-tags were cut on to enhance a data sample in b-events before running the main analysis e.g. to reduce the combinatorial background in the reconstruction of b-hadron states.
- Cutting on the b-tag and simply counting the number of events passing the cut selection e.g. as in the measurement of R_b.
- Using the whole tagging quantity distribution to fit for a parameter of interest e.g. the lifetime of a b-hadron state.
- A combination of b-tagging, to gain an enriched sample, and fitting for some parameter of interest from the distribution of another quantity.

During the LEP period, tags for b-quarks reached a remarkable level of performance culminating in the development of inclusive tags not only for b-quarks but also for tagging a particular b-hadron type ('flavour tagging') where the presence of a b or $\bar{b}$ quark in a hadron is tagged. These more sophisticated tagging methods are described in Chap. 7. In this chapter we concentrate on the more standard b-tags developed during the course of LEP, starting with lepton tags and ending with the development of combined tagging methods. Often, the price paid for an inclusive analysis is a greater dependence on simulation to model the various physics contributions to the distribution of an inclusive variable. The driving force behind the development of b-tags is the efficient rejection of backgrounds and this chapter ends with some examples of what can happen if the background to $Z^0 \rightarrow b\bar{b}$ events is not modelled precisely enough.

4.1 Lepton Tags

At the start of LEP, before fully functioning silicon vertex detectors came on the scene lepton tagging and, to a lesser extent, event shape variables (see below) were the most widespread methods of identifying b-quarks. Tags from these sources were

G.J. Barker, *b-Quark Physics with the LEP Collider*, STMP 236, 57–109,
DOI 10.1007/978-3-642-05279-8_4, © Springer-Verlag Berlin Heidelberg 2010

well known from low energy e^+e^- experiments and algorithms that had been successfully applied at PEP and PETRA were among the first to be used on LEP data.

About 20% of b-quarks decay semi-leptonically to electrons and muons. The hard fragmentation characteristic of heavy quarks (which is evident from Fig 7.13) and the lower number of decay particles in semi-leptonic decay (on average), mean that the lepton momentum spectrum is biased to large values. In addition, the high mass of the decaying b-quark imparts a large momentum component to the leptons transverse to the b-quark direction. The result is that about half of all the b-quark semi-leptonic decays at LEP produced a lepton with $p \gtrsim 3\,\mathrm{GeV}/c$ and $p_t \gtrsim 1\,\mathrm{GeV}/c$. An example of a p_t spectrum for muons is shown in Fig. 4.1 and illustrates that at large values, the spectrum is dominated by direct b-quark decays. When used as a b-tag, samples that are 70–90% pure in $Z^0 \to b\bar{b}$ events can be attained by cutting on p and/or p_t, with a selection efficiency of about 5% (with respect to the initial number of leptons from b-decay).

Although lepton identification is experimentally relatively simple (particularly for muons), there are some important restrictions and disadvantages linked to this method: The semi-leptonic branching ratio, of approximately 10% strongly restricts the overall tagging efficiency achievable. In addition, the uncertainity on the branching ratio becomes an irreducible source of systematic uncertainty on the tagging efficiency. A major hurdle to precision measurements with lepton spectra, is unravelling the various components from the direct $b \to \ell^-$ process we are interested in. Table 4.1 lists all the different sources of leptons in b-events that can contribute to the lepton momentum spectrum. The background is rather complicated – photon conversions, π and K decay all produce leptons. In addition fake muons can be produced from hadrons that do not interact in the material ('sail-through') or from hadrons whose interaction products penetrate the material ('punch-through').

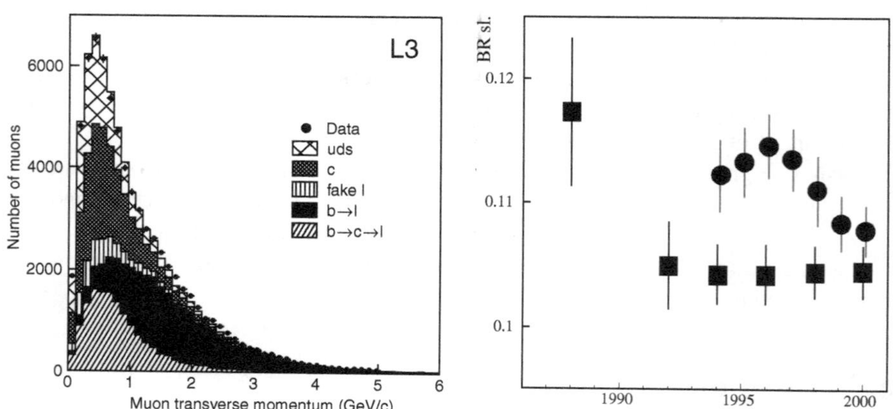

Fig. 4.1 (left) The p_t spectrum of muon candidates reconstructed by the L3 collaboration. (right) The variation in time of measurements of $BR(b \to \ell^+ \nu_\ell X)$ from LEP (*circles*) and from experiments running at the $\Upsilon(4s)$ (*squares*)

Table 4.1 Sources of leptons in b-events

Direct b decay	$b \to \ell^-$
Cascade b decay	$b \to c \to \ell^+$
$b \to cW$, $W \to \bar{c}s$	$b \to \bar{c} \to \ell^-$
$B^0 - \bar{B}^0$ mixing	$b \to \bar{b} \to \ell^+$
Tau decay	$b \to \tau^- \to \ell^-$
Backgrounds	$c \to \ell^+$, fake $\ell^\pm$, $\pi^\pm$, $K^\pm$, $\gamma \to e^\pm$

These reasons meant that lepton tags were only used in early LEP b-physics publications but continued to play an important role when used in combination with other independent techniques.

4.1.1 LEP Measurements of $BR(b \to \ell^- \bar{\nu}_\ell X)$

The impact of these issues is illustrated by early measurements of $BR(b \to \ell^- \bar{\nu}_\ell X)$ from LEP. This analysis involves fitting the proportion of essentially three different components; direct ($b \to \ell^-$), cascade $b \to c \to \ell^+$ and background, from the measured momentum distribution. The similar shapes of the direct and cascade components mean that it is not possible to unfold them from the data and instead they must be taken from models (e.g. ISGW [1]) that predict $b \to \ell$ and $c \to \ell$ spectra. Unfortunately the theoretical uncertainties associated with these models is relatively large (especially for the cascade component which is not directly modelled) and in the early years of LEP were probably underestimated. As tagging techniques progressed it became possible to clearly separate the direct and cascade lepton spectra contributions and unfold their shapes directly from the data with minimal model dependence. This is the main reason why, in Fig. 4.1 (right), the measurements of $BR(b \to \ell^- \bar{\nu}_\ell X)$ from LEP are significantly higher pre-1996 compared to later years. The same figure also displays results from the CLEO and ARGUS experiments operating at the $\Upsilon(4s)$ resonance, which interestingly were significantly lower than the LEP results and very stable over time. This is explained by the fact that these experiments were always able to measure the component shapes from the data simply because at the $\Upsilon(4s)$ the B is at rest and the direct and cascade spectra are very different. The improvements in tagging techniques that finally brought the semi-leptonic branching ratio measurements from LEP under control are discussed further in Chap. 7 together with the current status of this measurement.

4.2 Event Shapes

The *shape* of an event is a measure of how longitudinal and transverse momentum are shared among the various particles that make up the event. As was the case for semi-leptonic decay, the relatively high b-quark mass will lead to the mean transverse momentum content of b-jets being higher than for light-quark jets. In addition, the mean number of B-hadron decay particles that the available momentum is shared between is relatively high ($\sim$ five), resulting in the mean momentum of $Z^0 \to b\bar{b}$ events being slightly lower than for u, d, s and c events. This combination of slightly higher mean transverse momentum coupled with slightly lower mean momentum leads to b-jets being somewhat less collimated, or more isotropic than light-quark jets.

One way of quantifying this difference in shape is through the variables thrust and sphericity. The thrust axis is defined to be the direction that maximises the thrust value T of a group of particles $i = 1, \ldots, N$ where,

$$T = \frac{\sum_{i=1}^{N} |p_l^i|}{\sum_{i=1}^{N} |p^i|} \quad \left(\frac{1}{2} \leq T \leq 1\right), \tag{4.1}$$

and p_l is the component of particle momentum along the thrust axis. It follows that an extreme two-jet event would have $T = 1$. The sphericity axis is the direction along which the sphericity value S is minimised where,

$$S = \frac{3\sum_{i=1}^{N} {p_t^i}^2}{2\sum_{i=1}^{N} {p^i}^2} \quad (0 \leq S \leq 1) \tag{4.2}$$

and p_t is the component of particle momentum transverse to the sphericity axis. For extreme two-jet events $S = 0$ and S tends to 1 as an event becomes more isotropically distributed. Jets in b-events are more likely on-average to be associated with larger values of sphericity, and hence correspondingly smaller values of thrust, than the average c or light-quark event.

Distributions of thrust and sphericity, comparing b-events to the lighter flavours, are shown in Figs. 4.2(a) and (b) respectively. Although the distributions show a systematically different shape, the difference is minimal and the main reasons for this are not difficult to find: at LEP energies of $W \sim 90$ GeV the distinction between the decay of a 'heavy' and 'light' quark diminishes with respect to the situation at PETRA energies of $W \sim 30$ GeV. The larger boosts at LEP makes all jets appear more collimated or 'pencil-like'.

In addition to the parent quark mass, there are two further sources of jet p_t which are bad for b-tagging because they act to 'smear-out' the quark mass contributions to the p_t. The first is quark and gluon fragmentation which contributes on-average 0.3 GeV/c per track of p_t (relative to the parent quark direction) independent of the initial energy. This contribution is largely irreducible. The second source is gluon Bremsstrahlung from quarks and it is the sensitivity of sphericity and thrust to such QCD effects that are largely responsible for the long tails seen in their distributions (Figs. 4.2). This effect can in principle be reduced by looking in the direction out of the $e^+e^- \to q\bar{q}g$ event plane.[1]

[1] Since the process is observed in a frame of reference that is also the centre-of-mass frame, the net momentum must vanish. Hence the gluon must appear in the same plane as the two quarks and the resulting system of jets will then largely retain this planar topology.

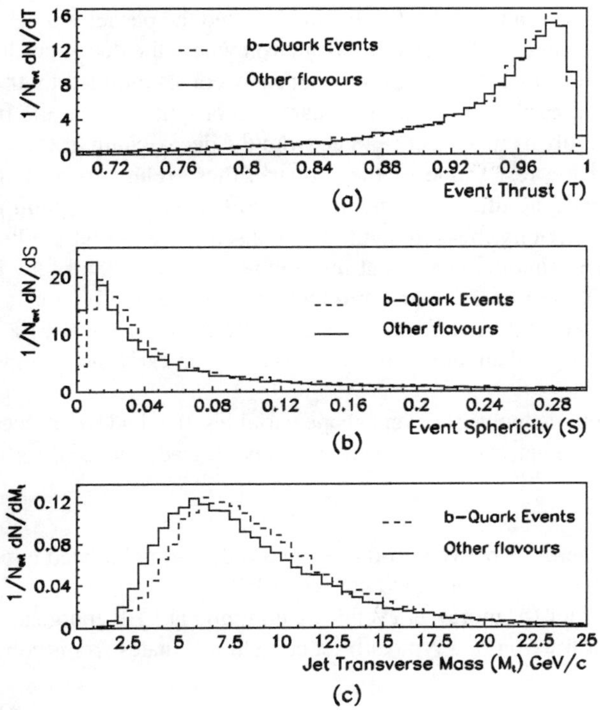

Fig. 4.2 (**a**) Event thrust, (**b**) event sphericity and (**c**) jet transverse mass distributions comparing b-quark events to c and light-quark events. From [2]

With these considerations in mind the variable *transverse jet mass* (M_t) has been studied where,

$$M_t = \frac{2E_{\text{beam}}}{E_{\text{vis}}} \sum_{i=1}^{N} \left| p_i^{\text{out}} \right| . \tag{4.3}$$

Here E_{beam} is the colliding beam energy and E_{vis} is the total visible event energy in charged particles. The quantity p_i^{out} is the component of particle momentum out of the plane defined by the thrust vector and the orthogonal axis that points in the direction that maximises the transverse momentum content of the event $\sum p_t$ with p_t calculated with respect to this axis. A distribution of M_t is presented in Fig. 4.2(c) where b-events are compared to the lighter flavours. Although, as expected, the flavour-discriminating power of M_t is found to be superior to sphericity or thrust tagging alone, the performance is not at the same level as was seen at e.g. PETRA where M_t was successfully used as a b-tag [3]. A reason for this is the increased frequency of events with four (or more) jets for which the planar three-jet topology is no longer true. These events will in general populate the tail of the M_t distribution and will act to blur the p_t^{out} contribution coming only from the quark mass.

Another tagging technique that had been used successfully at lower energies (e.g. at TASSO [4]) was the concept of *boosted* sphericity. The idea here is to calculate the sphericity as given by (4.2) after having boosted the particles in a hemisphere or jet into the rest frame of the decaying B-hadron where the decay products show the greatest isotropy. At LEP energies this picture becomes clouded by the fact that up to $\sim 40\%$ of the particle content of b-quark jets originates from the fragmentation process i.e. roughly as many as originate from the decay chain of the weakly decaying B-state. In the rest frame of the b-quark, these relatively soft fragmentation particles can have significant momenta back to the interaction point and so act to lower the jet sphericity measurement. The consequence is that the boost resulting in the best discrimination is not that into the rest frame of the b-quark, as was the case at TASSO, but rather a boost into the rest frame of the b-jet as a whole. Such complications reduced the effectiveness of boosted sphericity at LEP although the performance was good enough to form the b-tag for an early measurements of R_b by DELPHI [5].

In conclusion, the power of event shape variables, that had been successfully used for b-physics at lower energies, was found to be degraded at LEP for the following reasons:

- the mass difference between b-quarks and u,d,s,c-quarks becomes insignificant compared to the huge collision energy at LEP,
- non-planar, four (or more) jet events are common at LEP energies,
- a substantial fraction of a typical b-jet content originates from the fragmentation process.

In later developments at LEP, event shape variables were used successfully in conjunction with other methods, for instance as one of the many inputs of an artificial neural network (see Sect. 4.5).

4.3 D* Reconstruction

The reconstruction of D^{+*} mesons to tag B-decays has traditionally been a popular experimental method because of the relative ease with which the decay $D^{+*} \to D^0\pi^+$ can be reconstructed. The trick, first used by the MARK II Collaboration [6], is to reconstruct the mass difference $\Delta m = m\left(D^{+*}\right) - m\left(D^0\right)$ where D^* mesons appear as a very narrow signal peak on top of a low background from which a clean sample can be easily isolated. This situation arises because the nominal mass difference $\Delta m = 145.5$ MeV is very close to the kinematic threshold for pion production of 139.6 MeV. The pion from D^{+*} decay is therefore produced with a small momentum in the D^{+*} rest frame and is termed the 'slow pion', often labelled as π_*^+. Consequently any contribution to the uncertainty on Δm from the slow pion momentum determination is small and the D^{+*} Δm-peak will be reconstructed with good resolution. The significance of the signal peak can be very high since the decay occurs close to the kinematic threshold and leaves very little phase space for combinatorial background events to populate.

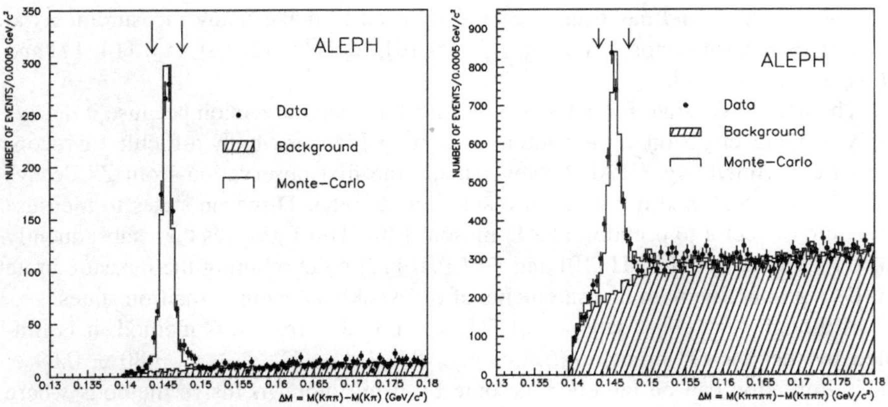

Fig. 4.3 Reconstructed D* signals from ALEPH based on fully reconstructed D^0 states in the channels (*left*) $D^0 \to K^-\pi^+$ and (*right*) $D^0 \to K^-\pi^+\pi^+\pi^-$ respectively. *Arrows* indicate signal isolation cuts in the analysis. From [7]

Results from ALEPH [7] are presented in Fig. 4.3 which shows that samples of D^{+*} can be isolated with essentially no background in the channel $D^0 \to K^-\pi^+$. Figure 4.3 also shows a clear signal from the channel $D^0 \to K^-\pi^+\pi^+\pi^-$ albeit with larger combinatorial background.

To produce such signals, an analysis would typically plot the invariant mass of all combinations of two-tracks or four-tracks with the correct charge assignments to be candidates for $K^-\pi^+$ or $K^-\pi^+\pi^+\pi^-$ respectively. Each track is assigned the assumed mass hypothesis i.e. m_K or m_π in the calculation of the invariant mass. Only those combinations that fall within a narrow mass window around the nominal D^0 mass are retained to be combined with low momentum pions (π^+_* candidates) and the mass of this combination is checked for compatibility with the nominal D^{+*} mass. Spurious track combinations that add combinatorial background under the signal mass peak can be reduced by demanding that kaon candidates pass some kaon identification criteria of the type discussed in Sect. 3.2.2. Another effective background rejection method is to form the angle θ^* between the D^0 flight direction and the kaon direction in the D^0 rest frame. Since the D^0 is a scalar particle, this distribution should be flat (in $\cos\theta^*$) for genuine $D^0 \to K^-\pi^+$ decays but strongly peaked in the backward direction for background. An illustration of this idea for the case of b-hadrons can be seen in Fig. 7.3. A reasonable background rejection cut would be $\cos\theta^* > -0.9$.

Good signals have also been routinely extracted using the channels, $D^0 \to K^-\pi^+\pi^0$ and $D^0 \to K^-\ell^+\nu_\ell$, although in these cases the resolution is degraded due to undetected neutral pions and neutrinos.

To good approximation, the resulting sample of D^{+*} candidates originates almost entirely from either $Z^0 \to c\bar{c}$ events or from b-hadron decays – with roughly equal rates. To help separate the two contributions, b-tag variables can be used and/or the fact that D^{+*} mesons from c-decay have a higher momentum than those from b-decay. Clearly this technique can just as easily be applied to isolate

$Z^0 \rightarrow c\bar{c}$ events and has been successfully used to make many measurements at LEP in the charm sector including: A^{c}_{FB} [8–10], R_c [11] [12, 13], $\langle x_c \rangle$ [14–17] and $P\,(g \rightarrow c\bar{c})$ [18, 19].

The neutral D^* state is far less useful than the charged version because it decays only via the emission of a photon or a π^0 which are both difficult to reconstruct experimentally. OPAL however made the first observation from Z^0 decays of D^{0*} and combined it with their results on the other D-meson states to measure the ratio of vector to pseudoscalar D-mesons [20]. This ratio was then subsequently also measured by ALEPH [19] and DELPHI [12] by combining the measured rate of D^{+*} production with measurements of the weakly decaying c-hadron states.

Although very clean samples of D^{+*} are possible from this method, it is limited by the low D^0 branching ratios e.g. $BR\left(D^0 \rightarrow K^-\pi^+\right) = 3.80 \pm 0.09\%$. This limitation can be largely overcome by using more inclusive methods where no attempt is made to exclusively reconstruct the D^0. The DELPHI collaboration first developed this idea [21] in order to boost the reconstruction efficiency in the important channel $\bar{B}^0_d \rightarrow D^{*+}\ell^-\bar{\nu}_\ell$ (which is described in more detail in the next section). The procedure involves first estimating the D^0 momentum by e.g. assigning a weight to each particle in a b-jet candidate, proportional to the probability that the

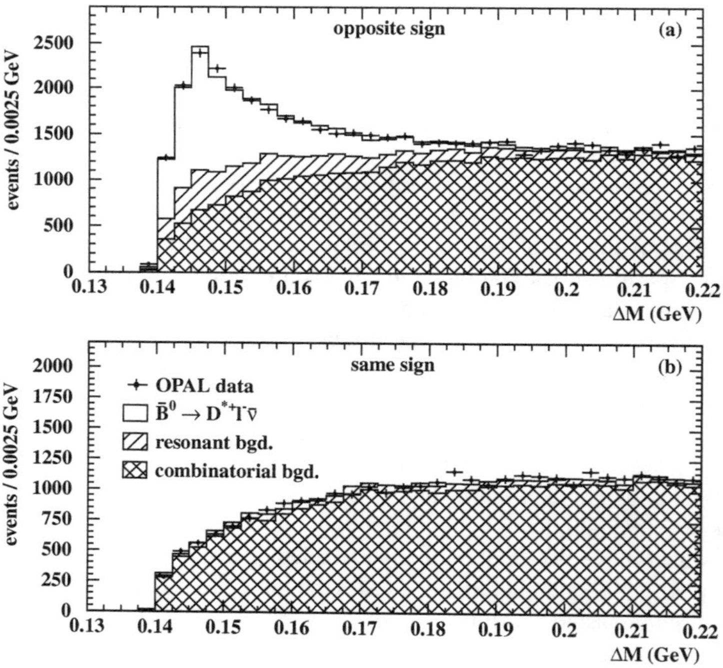

Fig. 4.4 Reconstructed $\Delta M = \left(D^{+*}\right) - m\left(D^0\right)$ distributions in OPAL data for (**a**) opposite sign and (**b**) same sign lepton-π_* combinations. Shown are the data (*points*) and the simulation contributions (*histograms*) where the resonant background is made up mainly of $\bar{B}^- \rightarrow D^{+*}\pi^-\ell^-\bar{\nu}_\ell$, $\bar{B}^0 \rightarrow D^{+*}\pi^0\ell^-\bar{\nu}_\ell$ and $\bar{B}_s \rightarrow D^{+*}K^0\ell^-\bar{\nu}_\ell$

particle is the daughter of a D^0 decay. Particles can then be combined in order of descending weight value until the combined invariant mass crosses the D-meson mass threshold. Next, for jets that contain a high p_t lepton, each charged particle in the jet is considered in-turn to be the slow pion candidate provided that it has low enough p_t with respect to the D^0 direction. If the slow pion was previously included in the D^0 momentum definition, it is removed and the D^0 momentum and energy recalculated. The final selection can then be made based on the mass difference Δm between D^{+*} and D^0 in a similar way as for exclusive D^{+*} reconstruction. Figure 4.4 shows the result of such an analysis from OPAL [22] where the data has been split into a sample where the lepton and slow pion have opposite electric charge, as is expected from the decay chain, and a sample where the charges are the same. This division is useful since the same-sign sample provides a good description of the combinatorial background and the signal all concentrates at low Δm values in the opposite-sign distribution. The fact that the inclusive approach can not match the signal to noise performance attained when the D^0 is fully reconstructed (compare Fig. 4.3 and Fig. 4.4), can be offset by the much larger samples of $\bar{B}^0_d \to D^{*+}\ell^-\bar{\nu}_\ell$ decays that can be obtained.

4.3.1 $\bar{B}^0_d \to D^{*+}\ell^-\bar{\nu}_\ell$ and $|V_{cb}|$

The exclusive channel $\bar{B}^0_d \to D^{*+}\ell^-\bar{\nu}_\ell$ was used extensively by the LEP experiments, primarily for the extraction of $|V_{cb}|$ but also to measure the B^0_d meson lifetime [22, 23] and to study $B^0_d - \bar{B}^0_d$ oscillations [22, 24].

As already noted, the reconstruction efficiency is good due to the presence of a high p_t lepton and D^{+*} meson. Add in the fact that the branching ratio of $5.53 \pm 0.23\%$ [25] is also the largest of all B^0 decay modes and $\bar{B}^0_d \to D^{*+}\ell^-\bar{\nu}_\ell$ really earns the title of an experimental 'golden' channel. The channel is also very interesting from a theoretical viewpoint since HQET studies of this mode [26–29], have revealed it to be largely free of the hadronic uncertainties associated with the binding of quarks and gluons into hadrons. The decay rate is parameterised as a function of the variable ω defined as the scalar product of the four-velocities of the D^{*+} and $\bar{B}^0$ mesons. This is related to the square of the four-momentum transfer q^2 from the $\bar{B}^0$ to the $\ell^-\bar{\nu}$ system by,

$$\omega = v_{\bar{B}^0} \cdot v_{D^{+*}} = \frac{E_{D^{+*}}}{m_{D^{+*}}} = \frac{m^2_{D^{+*}} + m^2_{\bar{B}^0} - q^2}{2m_{\bar{B}^0} m_{D^{+*}}} \tag{4.4}$$

where $E_{D^{+*}}$ denotes the recoil energy of the $E_{D^{+*}}$ in the $\bar{B}^0$ rest frame and $q^2 = (p_{\bar{B}^0} - p_{D^{+*}})^2$. Hence ω varies between a value of 1.0 when the D^{+*} is produced at rest in the $\bar{B}^0$ rest frame (i.e. the zero recoil points), to a value of about 1.5. Using HQET, the differential decay rate [30, 31] is given by,

$$\frac{d\Gamma\left(\bar{B}^0 \to D^{*+}\ell^-\bar{\nu}_\ell\right)}{d\omega} = \frac{1}{\tau_{B^0_d}} \frac{dBR\left(\bar{B}^0 \to D^{*+}\ell^-\bar{\nu}_\ell\right)}{d\omega} \tag{4.5}$$

$$= \mathcal{K}(\omega)\mathcal{F}^2(\omega)\,|V_{cb}|^2 \tag{4.6}$$

where $\mathcal{F}^2(\omega)$ is the hadronic form factor for this decay and $\mathcal{K}(\omega)$ is a known phase space term.

Although the shape of this form factor is unknown, its magnitude at zero recoil can be calculated using HQET which suggests that $|V_{cb}|$ can be extracted if a measurement of the differential decay rate can be made and extrapolated to the zero recoil point. The experimental challenge is to keep the reconstruction efficiency roughly constant all the way down to as close to the zero recoil point as possible so that the extrapolation is as accurate as possible. This strategy has been successfully applied by the LEP experiments and the differential rate spectrum from the OPAL analysis, based on the sample of Fig. 4.4, is shown in Fig. 4.5 [32].

In the heavy quark limit ($m_b \to \infty$), $\mathcal{F}(1)$ is the so-called Isgur-Wise function [27–29] which is normalised to a value of unity at the zero recoil point. Corrections to $\mathcal{F}(1)$ have been calculated [33] to account for the effects of finite quark

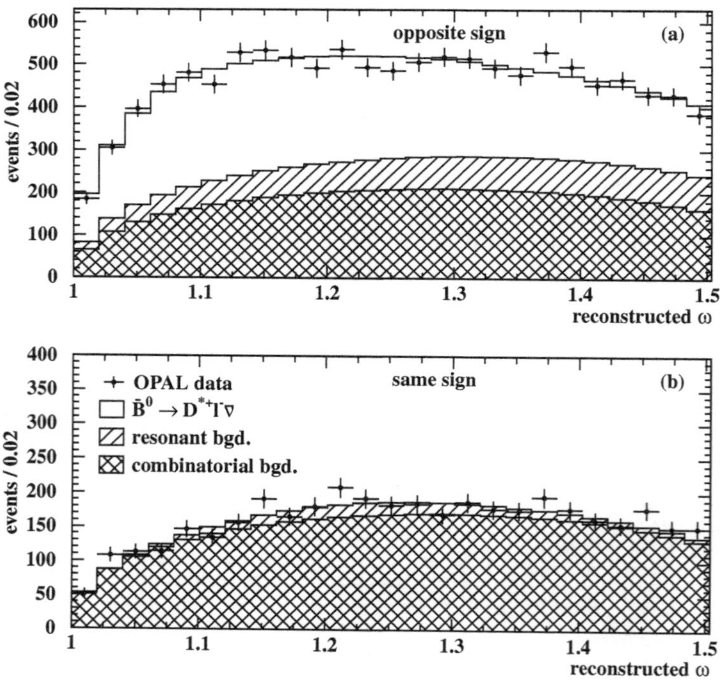

Fig. 4.5 Distributions of reconstructed ω in OPAL data for (**a**) opposite sign and (**b**) same sign lepton-π_* combinations. Shown are the data (*points*) and the fitted simulation contributions (*histograms*)

masses and non-perturbative QCD corrections. The main conclusion of this and the reason why the $\bar{B}_d^0 \to D^{*+}\ell^-\bar{\nu}_\ell$ channel is so theoretically 'clean', is that the leading corrections (i.e. to first order in Λ_{QCD}/m_b) are absent at the zero recoil point. This is not the case for the $\bar{B}_d^0 \to D^+\ell^-\bar{\nu}_\ell$ channel [34].

In practice experiments fit the measured differential spectrum $d\Gamma/d\omega$ to (4.5) using the HQET prediction for $\mathcal{F}(\omega)$, leaving the product $\mathcal{F}^2(1)\,|V_{cb}|^2$ and some HQET quantities as free parameters, convoluted with the expected experimental resolution. Table 4.2 collects together the LEP results on $|V_{cb}|$ using the exclusive channel $\bar{B}_d^0 \to D^{*+}\ell^-\bar{\nu}_\ell$. Results of similar precision also exist from CLEO, BELLE and BABAR, but this is clearly a measurement that will be dominated by the B-Factory experiments in the future.

It is worthwhile mentioning that the ALEPH analysis [35] uses the same method and the HQET predictions to extract the corresponding form factor for $\bar{B}_d^0 \to D^+\ell^-\bar{\nu}_\ell$ decays, which is commonly called $\mathcal{G}(\omega)$. With this measurement ALEPH were able to make an important direct test of the HQET prediction that the same form factor can describe the decay $\bar{B}_d^0 \to D^{*+}\ell^-\bar{\nu}_\ell$ and $\bar{B}_d^0 \to D^+\ell^-\bar{\nu}_\ell$. Figure 4.6 shows that the measured ratio $\mathcal{G}(\omega)/\mathcal{F}(\omega)$ is consistent with one across the full ω range. In addition, the measured value $\mathcal{G}(1)/\mathcal{F}(1) = 0.87 \pm 0.22(stat.) \pm 0.21(stat.)$ is consistent with the HQET prediction of $\mathcal{G}(1)/\mathcal{F}(1) = 1.08 \pm 0.06$ [38] and also with measurements of the same quantity by CLEO [39] and BELLE [40].

Table 4.2 $|V_{cb}|$ measurements from LEP based on exclusive reconstruction of $\bar{B}_d^0 \to D^{*+}\ell^-\bar{\nu}_\ell$. Numbers have been adjusted to use common inputs/assumptions (see Sect. 1.4.1)

Analysis	$|V_{cb}|\mathcal{F}(1)\ [10^{-3}]$	Remarks
ALEPH [35]	$33.6 \pm 2.1 \pm 1.6$	Exclusive $\bar{B}_d^0 \to D^{*+}\ell^-\bar{\nu}_\ell$ and $\bar{B}_d^0 \to D^+\ell^-\bar{\nu}_\ell$
OPAL [32]	$38.5 \pm 1.2 \pm 2.4$	Partial D^0 reconstruction
OPAL [32]	$39.1 \pm 1.6 \pm 2.3$	Exclusive D^0 reconstruction
DELPHI [36]	$36.9 \pm 1.4 \pm 2.5$	Partial D^0 reconstruction
DELPHI [37]	$38.7 \pm 1.8 \pm 2.2$	Exclusive D^0 reconstruction

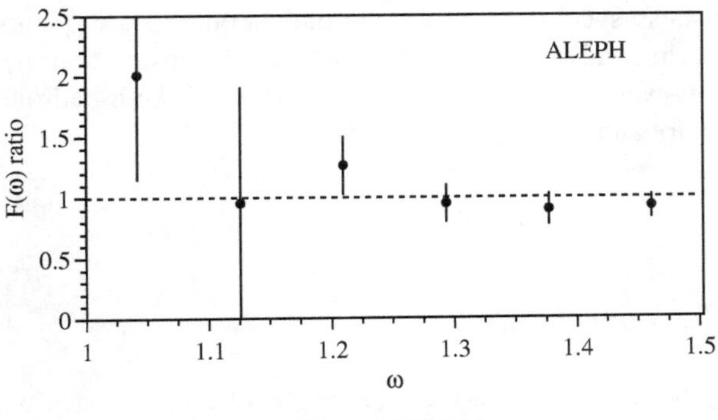

Fig. 4.6 The ratio $\mathcal{G}(\omega)/\mathcal{F}(\omega)$ as measured by ALEPH [35]

Before leaving the subject of $|V_{cb}|$, we note that in recent years the most accurate determinations have come from comparing the measured, inclusive, semi-leptonic branching ratio BR($b \rightarrow X_c \ell^- \bar{\nu}_\ell$) with the prediction based on the Operator Product Expansion formalism. The prediction depends on several non-perturbative QCD parameters, some of which have recently been experimentally determined via measurements of moments in B-meson semi-leptonic decay and $B \rightarrow X_s \gamma$ (see e.g. Chap. 3 of [41]).

This has resulted in a significant improvement in uncertainty and flipped the status of the analysis from being limited by theoretical uncertainties to being driven by advances in the moments measurements. This is in contrast to the situation for the exclusive analysis which is also currently limited by uncertainties in non-perturbative QCD quantities, but these cannot be measured an have to be theoretically predicted. Improvements in precision here will probably have to wait for developments in lattice QCD calculations of semi-leptonic form factors.

4.4 Lifetime Tagging

The LEP experiments, together with the SLC detector at SLAC, were the first collider experiments able to make measurements sensitive to the b-quark lifetime. This was made possible by two facts: (a) b-hadrons travelled an average distance of $\sim$ 3 mm before decaying, and (b) the use of silicon vertex detectors with point resolutions of the order 10 μm in both the $R - \phi$ and $R - z$ plane. The ability to reconstruct lifetime information helps enormously in b-tagging since, unlike lepton tagging or D* reconstruction, lifetime tagging can be applied to the whole data set and is not dependent on specific channels with small branching ratios. In fact, it turned out that the LEP detectors, in their final configurations were so well suited to lifetime reconstruction that methods of lifetime b-tagging eclipsed the performance of all other techniques.

Experimentally, the quantities that formed the basis for lifetime tagging, (lifetime-signed) impact parameter and decay length, were described in Sect. 3.1. Before going on to discuss specific lifetime tags built around these quantities, we now discuss the experimental issues necessary for their reconstruction: determining the jet-structure of b-events which provides an estimate of the b-hadron direction and finding the primary and secondary vertex points.

4.4.1 b-jet Reconstruction

Reconstructing the jet of particles (charged and neutral) associated with a b-hadron plays an important role in b-tagging in order to:

(a) Estimate the direction of the b-hadron trajectory. This is necessary to lifetime-sign the impact parameter or decay length and to calculate the p_t of a lepton with respect to the b-hadron direction.
(b) Reduce the combinatorial background when applying algorithms to reconstruct the b-hadron or its decay vertex position.

For b-tagging, the ideal jet finding algorithm would find jets which included all particles coming from a single b-hadron decay and excluded all other particles. This way, the sum of the four-vectors of particles in the jet would give precisely the energy and momentum of the parent b-hadron and any algorithm to reconstruct the mass or decay point of the parent could run just on the particles in the jet with zero background.

In the early days of LEP b-physics e.g. pre-1993, the jet-clustering algorithms used were dictated largely by whatever was the state-of-the-art for QCD studies at the time. Most widely used at this time was the JADE algorithm [42] which is an example of a *binary clustering* scheme. In these algorithms a initial list of particles, which could be e.g. all particles in an event that have passed some basic quality requirements, are considered as the initial set of clusters. A separation measure d_{ij} is defined between two clusters i and j as a function of their four-momenta, $p_{i,j} = \left(E_{i,j}, p_{i,j}\right)$. The two clusters with the smallest d_{ij}^2-value are then merged into one, provided that d_{ij} is less than some cut-off value d_{cut}. The four-momentum of the new cluster k is simply the sum $p_k = p_i + p_j$ and this combination procedure is repeated until all pairs of clusters have a separation above d_{cut}. Often, the scaled parameter $y_{ij} = d_{ij}^2/E_{\text{vis}}^2$ is preferred as the separation measure where E_{vis} is the total visible energy in the detector. Normalisation to E_{vis} provides a certain degree of cancellation of systematic errors between the numerator and denominator and in this case clustering is stopped above the cut-off value $y_{ij} = y_{\text{cut}}$.

The separation measure in the JADE algorithm is defined as,

$$d_{ij}^2 = 2E_iE_j\left(1 - \cos\theta_{ij}\right) \tag{4.7}$$

where E_i, E_j, are the respective energies of the two clusters in the pairing and θ_{ij} is the opening angle between their momentum vectors. d_{ij}^2 is closely related to the invariant mass of the pairing since,

$$m_{ij}^2 = \left(p_i + p_j\right)^2 = m_i^2 + m_j^2 + 2\left(E_iE_j - |p_i|\,|p_j|\cos\theta_{ij}\right). \tag{4.8}$$

Many variations of the original algorithm with different combination schemes called e.g. JADE-E,-E0,-p [43] have subsequently appeared, all of which are essentially equivalent from the point of view of b-physics. Within any particular scheme however, the choice of cutoff point (d_{cut} or y_{cut}) must be tuned and depends to some extent on the particular b-physics application. A trade-off exists between setting the cutoff small enough so that particles not associated with the b-hadron decay chain are rejected from the jet, and setting it large enough so that real b-hadron decay products do not start to be *split* between more than one reconstructed jet.

Simulation can be used to make studies of the optimal choice for b-hadron direction and/or energy reconstruction. In addition, studying the origin of the particles assigned to jets as a function of the cutoff gives information about the optimal choice to reduce combinatorial background for b-hadron reconstruction. Figure 4.7 shows the result of a study [2] to optimise the choice of y_{cut} when using JADE jets to define an initial track list from which secondary vertices were reconstructed. Defining per jet,

$$S = \frac{\text{number of tracks in jet from b hadron decay chain}}{\text{number possible}}$$

$$B = \frac{\text{number of background tracks in jet}}{\text{total number of tracks in jet}}$$

and forming the ratio S/B provides a quantity that decreases at low y_{cut} due to the onset of jet splitting and at high y_{cut} because after a point only background tracks are being added to the jet. Figure 4.7 shows that a maximum occurs at $y_{\mathrm{cut}} = 0.01$ and corresponds to the point at which the b-hadron content of the jet is maximised with respect to the background sources. Strong support for this conclusion is given by the fact that the resolution on the b-hadron direction as estimated by the jet axis also minimised at this value for y_{cut} at a value of $\Delta_\phi \simeq \Delta_\theta \simeq 50$ mrad.[2] These results were confirmed by similar independent studies [44].

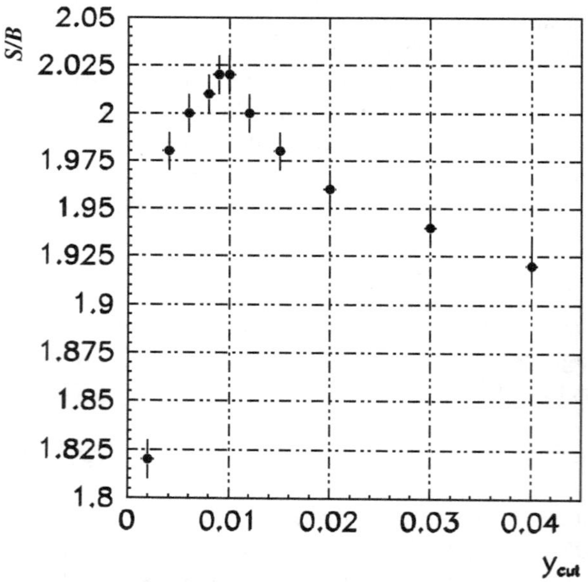

Fig. 4.7 The variation in S/B as a function of y_{cut} using JADE jet clustering

[2] As given by the width of a Gaussian fit to the core of the residuals distribution but which typically has non-Gaussian tails.

Invariant mass clustering was subsequently found to be not necessarily optimal for b-physics due to the realisation that particles tend to cluster closer in invariant mass when in a low momentum environment. This means that the JADE algorithm starts to cluster in the centre of the event among the fragmentation tracks, which carry little information about the parent b-hadron direction, and only afterwards spreads outwards to add in the faster particles which should be more important in representing the jet direction. Rather than clustering slow particles around the fast ones, invariant mass clustering will tend to cluster fast particles around slow ones and this has consequences for how accurately JADE-type algorithms can reconstruct b-hadron directions. The bias to soft particles also means that these algorithms are sensitive to details of fragmentation which ultimately limits their usefulness for QCD studies.

Cone algorithms represent an alternative jet clustering algorithm that do not suffer from the limitations of the invariant mass scheme. Such algorithms have traditionally been used at hadron colliders to define jets of energy deposits in calorimetry and were even used in the e^+e^- experiment PLUTO in 1980 [45]. It was not however until after around 1993 that they were applied to the LEP data. A cone algorithm defines a jet as a set of particles all lying within a cone drawn in space with the interaction point of the event as its apex. The size of the cone is parameterised by the half-angle R and the four-momentum of the jet is defined to be the sum of the four-momenta of the constituent particles. Therefore all particles with four-momenta p_i belonging to jet p_J must satisfy,

$$p_J = \sum_{i \,\epsilon\, \text{jet}} p_i \;\; ; \;\; \hat{p}_i \cdot \hat{p}_J \leq \cos(R) \;\; \forall\, i \,\epsilon\, \text{jet} \tag{4.9}$$

and the algorithm seeks to find as many such jets in an event as possible. Each jet satisfying this condition is required to have an energy exceeding a cut-off value ε and finally some procedure is executed in order to disentangle cones that overlap in space meaning that some particles could be assigned simultaneously to two (or more) jets.

A direct comparison between the JADE-E0 and a cone algorithm was performed on OPAL data [46]. Figures 4.8(a) and (b) display the mean of the azimuthal angle residual between the jet axis and the true b-hadron direction for the JADE-E0 and cone algorithms respectively.[3] The cone algorithm was run with ε set to 5 GeV. The results confirmed that the cone algorithm could indeed reproduce the b-hadron direction rather better than the JADE-E0 scheme and that the precision improved as the resolution of the jets gets finer i.e. as they become more pencil-like. Studies also found that particles omitted from b-hadron jets by the cone algorithm tended to be softer than those omitted from JADE-E0 jets. This is consistent with the picture of

[3] The best angular resolution quoted here for the JADE algorithm is larger than the 50 mrad cited earlier because these numbers derive from the mean of the absolute value of the residuals which is influenced more by outliers in the non-Gaussian tails.

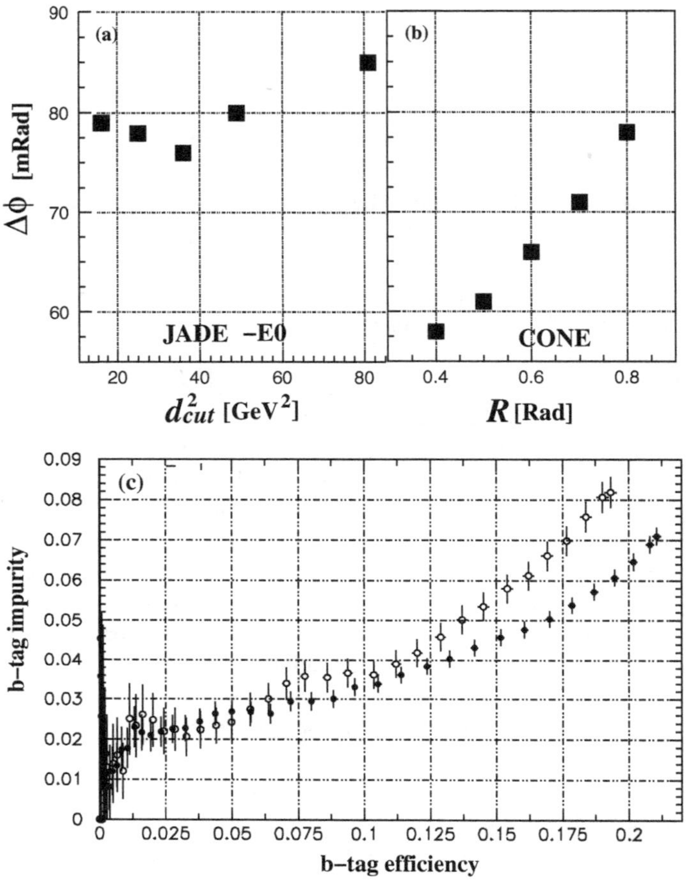

Fig. 4.8 $\Delta\phi$ as a function of scanning different values for (**a**) d^2_{cut} in the JADE-E0 algorithm (**b**) R in a cone jet finder and (**c**) shows the impurity from a secondary vertex b-tag as a function of tagging efficiency based on jets found by the cone algorithm $R = 0.55$ Rad,$\varepsilon = 5$ GeV(*closed circles*) and by the JADE-E0 scheme $d_{\mathrm{cut}} = 7$ GeV/c (*open circles*)

the high momentum particles, which follow closely the b-hadron direction, sitting at the heart of the jet with softer particles surrounding them. This is in contrast to the situation for the JADE-E0 algorithm where the presence of softer, mainly fragmentation, tracks sitting at the heart of the jet mean that the angular resolution begins to degrade as the jet resolution gets finer. Figure 4.8(c) shows how the performance of a decay length b-tag was affected by the choice of jet algorithm. Secondary vertices were reconstructed from the charged particles in jets and the decay length between the primary and secondary vertex was cut on to tag b-hadrons. Plotted is the impurity (i.e. 1-(fraction of tagged vertices)) against the tagging efficiency for the two algorithms and shows that the performance from the cone-based tag is superior. This result supports the premise that the cone algorithm acts to reject some of the softer particles not directly associated with the b-hadron decay chain and hence improve secondary vertex reconstruction by reducing the combinatorial background that the vertexing algorithms have to deal with. A later study by ALEPH [47] based on semi-leptonic decays, saw a 10% improvement in both the selection of b-decay tracks and the rejection of background fragmentation tracks compared to a JADE-E0 algorithm. Studies of this kind concluded that optimal settings for the cone algorithm parameters were in the range $R = 0.4 - 0.7$ Rad and $\varepsilon = 5$ Gev/c^2.

As a result of these studies, the use of invariant mass clustering for b-physics was replaced by either cone algorithms or by clustering algorithms based on transverse momentum such as LUCLUS [48] where the separation measure is,

$$d_{ij}^2 = \frac{2\,|p_i|^2\,|p_j|^2\,(1-\cos\theta_{ij})}{\left(|p_i|^2\,|p_j|^2\right)^2}. \tag{4.10}$$

DELPHI found that optimal values for the clustering cut-off for b-physics depended on whether an accurate b-hadron direction was required, e.g. as in [49] where d_{cut} was set to 3.5 Gev/c, or whether a jet definition to beat down combinatorial background was more important, as in inclusive b-hadron reconstruction [50] where a value of $d_{\text{cut}} \sim 5.0$ Gev/c was preferred. Studies have shown that p_t-based clustering algorithms also outperform the JADE algorithm in reconstructing jet energy and directions accurately and advances in the understanding of pQCD have spawned a whole new generation of p_t-based algorithms with names such as DURHAM, GENEVA and CAMBRIDGE (see e.g. [43] for further details) which have replaced the JADE algorithm as the new standard for QCD studies.

For the task of b-hadron direction reconstruction, use of a jet axis can be drastically improved on by applying techniques which attempt to directly evaluate, on a particle-by-particle basis, how likely it is that the particle originated from the decay of the b-hadron. We discuss here briefly two examples but details of these more advanced techniques are described in Chap. 7. If the probability (W_i) that a particle i originates from the b-hadron decay chain can be reconstructed, an estimate of the b-hadron flight direction is given by the weighted sum p_{B}^{WS} over all tracks in a hemisphere: $p_{\text{B}}^{WS} = \sum_i p_i \cdot W_i$. The probability W_i can also be used to select tracks from which the b-hadron decay vertex point can be reconstructed with good precision. The b-hadron direction p_{B}^{SV} can then be approximated as the vector joining the event primary vertex position and the decay secondary vertex: $p_{\text{B}}^{SV} = SV - PV$. Figure 4.9 shows the resolution on the b-hadron direction in ϕ from the two variables p_{B}^{WS} and p_{B}^{SV} as a function of cutting on the reconstructed decay length. The results show that the direction resolution attained by these techniques is better than from use of the jet axis and also that it improves in samples where the decay lengths are large. This is particularly true when using the primary to secondary vector and follows because, in general, the precision with which the secondary vertex can be reconstructed improves as the decay length increases. This observation tends to get washed-out when considering the direction resolution in polar angle θ since experiments tend to have a worse spatial resolution in the z-plane

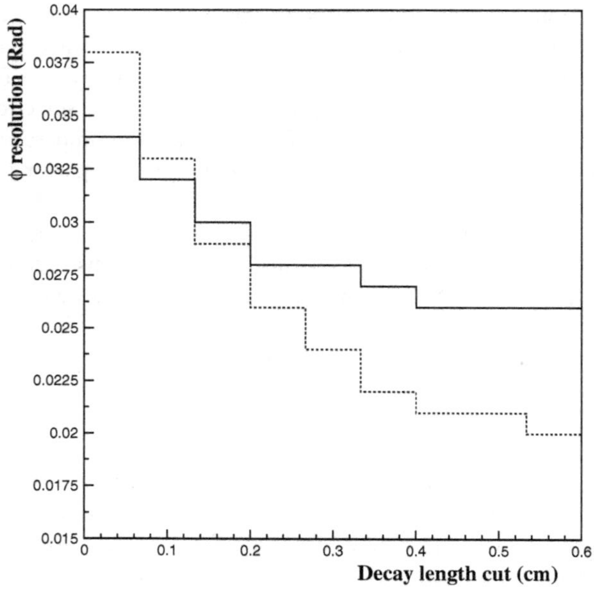

Fig. 4.9 A study of the ϕ resolution on the b-hadron direction attained by the p_{B}^{WS} (*solid line*) and p_{B}^{SV} (*dashed line*) variables as a function of cutting on the reconstructed decay length (Based on a Monte Carlo sample of b-hadron decays interfaced to the DELPHI simulation)

compared to $r - \phi$. Depending on the detector therefore, the resolution in θ may or may not be better from the primary to secondary vector method. For the case of b-hadrons that decay semi-leptonically it is noticed that the direction resolution can degrade badly (by 20–30%) from the momentum sum method due to the undetected neutrino. The effect is much smaller (a few percent) from the primary to secondary vector variable, since missing one particle from a vertex fit is unimportant providing that the remaining particles in the vertex are really from the b-hadron decay.

In summary therefore, the conclusion of the work to optimise jet clustering for b-physics was that a cone-type of algorithm was best suited to the physics of b-jet reconstruction in the LEP environment. If however a faithful reconstruction of the b-hadron direction is required and if an efficient method of selecting tracks associated with the b-hadron chain is available, one can do better by using the vector joining the primary to secondary vertices.

4.4.2 Primary Vertex

On an event-by-event basis the primary vertex at LEP could be located anywhere within the zone of intersection of the two colliding beams. This region, due to the LEP machine optics, took the form of a highly flattened ellipse in the $x - y$ plane (see Fig. 3.1) with dimensions $\sigma_x \lesssim 150\ \mu$m, $\sigma_y < 10\ \mu$m and extending several millimetres along the beam direction. In principle, the most precise determination of the primary vertex, at least in the x and z directions, is given by fitting the vertex for each event. In practice, selecting tracks for the fit that come from the primary vertex is a difficult task and so the first attempts to reconstruct the primary vertex by the LEP experiments concentrated on determining the average location of the central point of the beam ellipse. This point is commonly referred to as the 'beam-spot'.

4.4.2.1 Beam-spot

Two main approaches were developed to find the beam-spot. The first, as used by e.g. OPAL, involved accumulating charged particle tracks over several events and fitting them to a common vertex position when a total of about one hundred tracks had been collected. In order to minimise the effects of including tracks originating from secondary vertices, the tracks selected must have impact parameters less than an upper limit and the track contributing the largest χ^2 to the fit is rejected, and the fit made again, until all tracks make χ^2 contributions of less than ten units. The method then forms a weighted average of thirty such vertices occurring sequentially in time but respecting the boarders between machine running periods. The error on the beam-spot from this method is dominated by the size of the beam ellipse. To estimate the error on the x coordinate, an impact parameter distribution (d) for track trajectories along the x-axis is formed. Assuming the error on the average position is negligible compared to the beam ellipse spread in x, the measured impact parameter resolution from this class of track receives a contribution from the tracking σ_d and from the size of the beam ellipse in y which is tightly constrained. The σ_d found in this way is then subtracted off (in quadrature) from the width of the impact parameter distribution for tracks with trajectories following the y-axis to yield the size of the beam ellipse in x.

The second method involves averaging over primary vertices reconstructed event-by-event and can be illustrated by the method developed by DELPHI. Vertex fits from contiguous events, containing at least three tracks, with impact parameters less than 1 cm, are collected in time periods of about twenty minutes and respecting the boarders between machine running periods as before. These collections are then fitted to a common mean vertex position with associated errors. This scheme does not make as good use of the information available as the previous method to find the beam $\{x, y, z\}$ position since events with less than three accepted tracks are not used. It turns out however that the accuracy is adequate ($\sigma_x \sim 15\ \mu$m, $\sigma_y \sim 5\ \mu$m) and the method is much better optimised to find the size of the beam ellipse since the event-by-event information is retained. The ellipse size in y is not fitted since the intrinsic spread in y cannot be improved on. Figure 4.10 shows the beamspot position as reconstructed by this method. The entries at $y = 0$ correspond to fills for which there is no data. The beam is seen to be stable to within a few hundred microns during the five years from 1991 to 1995 but there are many significant deviations which are correlated with LEP machine running periods e.g the 1995 data can be

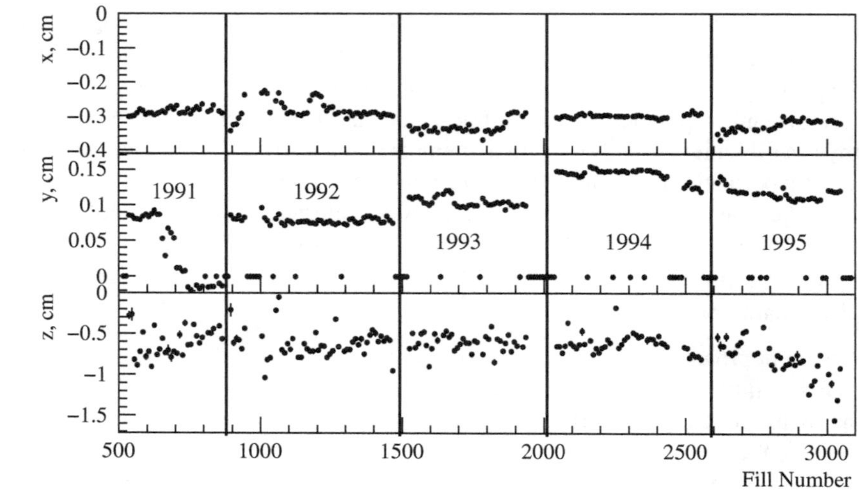

Fig. 4.10 The reconstructed position of the LEP beamspot as seen by DELPHI as a function of the year

seen to be less stable in z which is correlated with the start of LEP bunch train operation in preparation for LEP II.

The fitted size of the beam ellipse in x and z from this method is shown in Fig. 4.11. A significantly smaller spread in x is seen in 1992 which coincides with the machine operating at the beam-beam limit. This had the consequence that as the initial current was increased, so the focusing had to be relaxed.

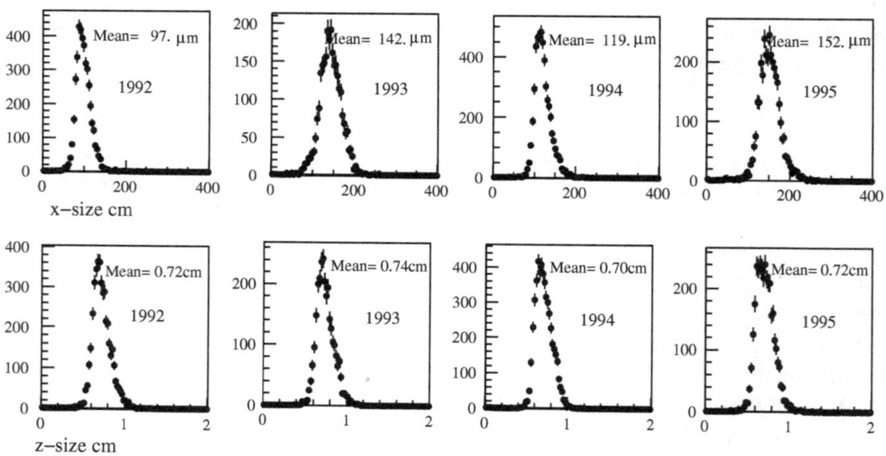

Fig. 4.11 The fitted size of the LEP beam ellipse in x and z as seen by DELPHI as a function of the year

4.4.2.2 Primary Vertex per Event

As already noted, fitting a primary vertex position on an event-by-event basis is difficult because of contamination of the track sample with the decay products of long-lived particles or with tracks from interactions with the detector material. Unfortunately these problems are worse in heavy flavour events, when often the multiplicity of tracks originating from the primary vertex is lower than that from the heavy flavour decays. To make matters worse, these primary tracks are frequently of low momentum which, due to multiple scattering, will give little position information when extrapolated back to the primary vertex. As experience with silicon vertex detector tracking grew, ways to select primary tracks reliably were developed and eventually, precise, full three-dimensional primary vertex positions were routinely reconstructed.

Algorithms typically would apply an initial track selection demanding the tracks carry some hits from the silicon detectors in the $x - y$ and z plane and that they are consistent with originating from the beam-spot position in terms of a cut on $\delta/\sigma(\delta)$ where δ is the closest approach of the track to the beam-spot and $\sigma(\delta)$ is the associated error. This group of N tracks would then be fitted to a common vertex, using the beam-spot and associated error as a constraint, and a $\chi^2(N)$ computed. In order to thoroughly reject secondary tracks from the fit, the next step involved removing each track i consecutively and repeating the fit to obtain corresponding $\chi^2(N_i)$ values. The track i giving the maximal difference $\Delta\chi^2 = \chi^2(N) - \chi^2(N-1)$ is rejected from the fit if this difference exceeds a threshold value $\Delta\chi^2(max)$ typically set to something like six. This procedure would then be repeated until there are no tracks with a χ^2 difference exceeding $\Delta\chi^2(\mathrm{max})$. Since the beam-spot is used as a constraint in the fit, it can also happen that all tracks are rejected and in this case the primary vertex coincides with the beam-spot and its covariance matrix corresponds to the beam-spot size. The fraction of such events is about 1% for Z^0 hadronic events. ALEPH's approach to removing the effects of secondary tracks is to first cluster the event into jets and then to project each track into the plane perpendicular to the jet it is associated with [51]. If the particle originated from the decay of a long-lived particle, this projection removes most of the effect of the lifetime. These projected tracks are then combined with the beamspot position to find the primary vertex point.

Representative values for the resolution attained by such algorithms in $Z^0 \rightarrow b\bar{b}$ events are ($\sigma_x = 60\,\mu\mathrm{m}, \sigma_z = 70\,\mu\mathrm{m}$) from DELPHI [52] and ($\sigma_x = 50\,\mu\mathrm{m}, \sigma_z = 60\,\mu\mathrm{m}$) from ALEPH [51], although the actual resolution achieved depends strongly on the number of tracks. Note that the precision on the y-coordinate is completely dominated by the beam-spot constraint ($\sigma_y \sim 10\,\mu\mathrm{m}$). The corresponding resolutions from DELPHI for $Z^0 \rightarrow$ u, d, s and $Z^0 \rightarrow c\bar{c}$ events are ($\sigma_x = 36\,\mu\mathrm{m}, \sigma_z = 43\,\mu\mathrm{m}$) and ($\sigma_x = 44\,\mu\mathrm{m}, \sigma_z = 50\,\mu\mathrm{m}$) respectively. The degraded performance in b-events compared to u,d,s and c-events is due to a lower track multiplicity from the primary vertex and to the increased risk of contamination of tracks from secondary decays. We note here that the corresponding numbers from OPAL [53] are ($\sigma_x = 50\,\mu\mathrm{m}, \sigma_z = 80\,\mu\mathrm{m}$) degrading to

($\sigma_x = 70\,\mu m, \sigma_z = 110\,\mu m$) when a primary vertex is fitted separately in each hemisphere of an event. This is understandable since the average track multiplicity is halved when dealing with event hemispheres but none-the-less, primary vertices constructed independently per hemisphere played an important role in reducing hemisphere efficiency correlations in many LEP analyses and is discussed further in Chap. 6.

4.4.3 Secondary Vertex

A good strategy with which to reconstruct the decay point of the weakly decaying b-hadron state (the secondary vertex in the event hemisphere) depends strongly on the physics objective. For example, if one were interested in reconstructing D_s^+ mesons in the channel $D_s^+ \rightarrow \phi\pi^+$ with $\phi \rightarrow K^-K^+$, a suitable strategy would be:

- Begin by reconstructing the ϕ. Select all opposite-sign track pairs that are consistent with a kaon signature in the detector (according to particle identification information).
- Fit the two tracks to a common vertex and calculate their combined invariant mass.
- Select those combinations with a small χ^2 for the vertex fit and with invariant mass close to the nominal ϕ mass, as ϕ candidates.
- Now the D_s^+ can be reconstructed by adding a third track, consistent with a pion signature, to the kaon candidates. Fit the tracks to a common vertex, require a small χ^2 and check the invariant mass distribution for a D_s^+ 'bump' at the characteristic mass.
- Reduce the number of wrong combinations further by checking that the angular distribution of the decay products is consistent with the decay of a parent state with a particular spin. In our example, the forward direction is favoured because of the P-wave nature of the decays and so a reasonable selection cut to make would be $|cos(\theta^*)| > 0.4$ where θ^* is the angle between the kaon direction in the ϕ rest frame and the ϕ boost direction in the D_s^+ rest frame.

This example illustrates two key points to successful secondary vertex reconstruction in a detector with finite spatial resolution: (i) the initial selection of tracks, and (ii) testing how well different combinations of tracks fit to a common vertex. Use of particle identification information, kinematical quantities such as invariant mass and knowledge of the underlying physics process to predict e.g. decay angles, allow the number of potential combinations of tracks to be drastically reduced. The χ^2 of the vertex fit can then be used as a tool to indicate how likely it is that a particular combination is correct.

The *exclusive* reconstruction of channels of this kind have been successfully used at LEP, particularly by the ALEPH Collaboration [54], to reconstruct not only $D_s^+ \rightarrow \phi\pi^+$ but also c-hadron samples in the channels $D^0 \rightarrow K^-\pi^+$, $D^+ \rightarrow$

$K^-\pi^+\pi^+$ and $\Lambda_c^+ \to pK^-\pi^+$. Exclusive reconstruction however has limited application, particularly for b-tagging, because of the small number of events that can be reconstructed in this way and the intrinsic inefficiency of the method. More important for LEP b-physics were the methods that addressed the thorny problem of *inclusive* vertex reconstruction. The task is to reconstruct the secondary vertices of weakly decaying b-hadrons without exact knowledge of the decay multiplicity or the type of particles the hadron decayed into. The topology of such events was illustrated in Fig. 1.3 where the primary, secondary and tertiary (the cascade c-hadron decay) vertices all occur within small distances of each other. Unless the spatial separation of the true vertices is significantly greater than the tracking resolution, ambiguities in associating tracks to vertices can easily occur i.e. it is not possible to uniquely assign a large fraction of the tracks to a single vertex. To improve this situation, the topological information has to be supplemented by criteria based on knowledge of the underlying physics.

An obvious place to start, is to make an initial track selection, rejecting those particles which are very unlikely to have come from the b-hadron weak decay. Even after a track selection procedure has been applied the track multiplicities in $Z^0 \to b\bar{b}$ events are frequently too high to make a search from all possible combinations a viable option. More efficient algorithms are therefore required with the emphasis on keeping the combinatorial search to a minimum. In addition, since most of the combinations tested will be wrong and mostly rejected by the algorithm, it is important to not calculate more information about a vertex than is absolutely necessary. For this reason most topological methods do not by default fit the vertex position by varying the track parameters within their measurement errors. Instead it is possible to get identical results by finding a point in space whose weighted distance sum from the measured tracks is minimal. This method has a computing time that typically rises linearly with the number of tracks in the fit n_t, compared to a rise with n_t^2 when re-fitting tracks. If later, the momentum of the tracks at the secondary vertex point is required e.g. in order to determine the invariant mass of the vertex, the transformation can be made after the vertex fit to only those vertices that are good candidates.

4.4.3.1 The Strip-Down and Build-Up Methods

These algorithms rely heavily on an initial pre-selection of tracks to remove the majority of tracks not originating from b-hadron decay while at the same time keeping a high efficiency for selecting the tracks of interest. Typical selection criteria include:

- Clustering the event tracks into jets, as described in Sect. 4.4.1, and only considering those tracks e.g. in the highest energy jets under the assumption that these are the most likely to contain a decaying b-hadron.
- Selecting tracks that are displaced from the primary vertex position based on e.g. their signed impact parameter or distance to the crossing of the track with the b-hadron flight direction.

- Tagging tracks likely to come from b-hadron decay based on kinematic and/or angular information e.g. track rapidity or helicity angle.
- Using particle identification measurements to select tracks often associated with the decay of b-hadrons e.g. leptons and kaons.
- Running algorithms to reconstruct the long-lived 'background' states, $K_s^0 \to \pi^+\pi^-$ and $\Lambda \to p\pi^-$ as well as photon conversions $\gamma \to e^+e^-$ and interaction vertices with detector material. These vertices can be readily identified with the techniques of exclusive reconstruction described above and the associated tracks removed from further consideration.

The *Strip-Down* method applies a similar ansatz to that already described in relation to primary vertex reconstruction: all tracks passing the selection are fitted to a common vertex and then, in an iterative loop, the largest contributor to the χ^2 is identified, stripped away (if its χ^2 contribution is deemed large enough) and the fit repeated for as many times as it takes for no tracks to pass the test for being stripped away. The OPAL collaboration has successfully applied such an algorithm in many b-physics analyses: A_{FB}^b, R_b and b-hadron lifetimes where the typical χ^2 contribution limit for stripping was set at four and vertices were required to contain at least three tracks. The efficiency for finding a secondary vertex using a strip-down algorithm applied to DELPHI $Z^0 \to b\bar{b}$ simulated events is shown in Fig. 4.12. To calculate the efficiency, a b-hadron vertex is flagged as reconstructed

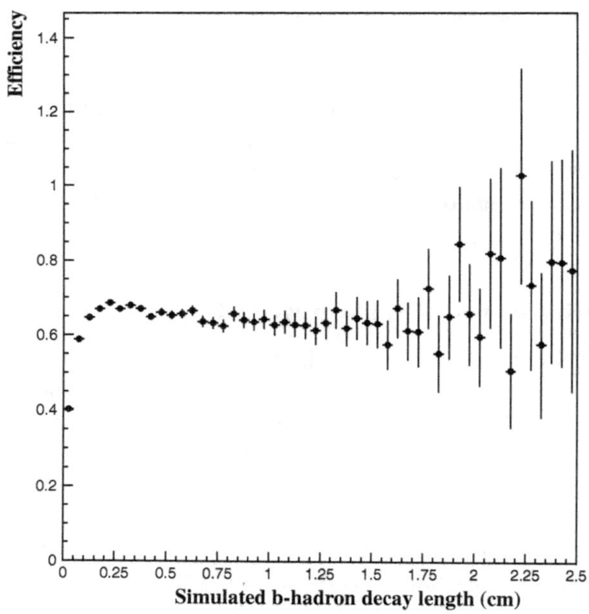

Fig. 4.12 The efficiency for finding a b-hadron secondary vertex as a function of the generated b-hadron decay length from a strip-down algorithm

if the reconstructed decay length agrees with the generated value to within three measurement standard deviations. Figure 4.12 shows a clear drop in efficiency at small decay lengths where discrimination of the b-hadron vertex from the primary vertex is difficult.

In contrast, the *Build-Up* approach begins with the reconstruction of all possible 'seed' vertices out of tracks that pass the pre-selection criteria. The next step involves rejecting,

- the more obvious spurious combinations by demanding all vertices have e.g. χ^2 probability $> 1\%$,
- any remaining K_s^0 and Λ decay vertices or material interaction vertices by demanding the decay length to the vertex is less than a few cm's,
- primary vertex candidates by demanding the decay length is larger than the measured error on the decay length.

The algorithm now attempts to build-up the seed vertex or vertices that survive by adding tracks that pass the selection cuts and keeping them permanently in the vertex definition if 'consistent' with originating from the seed. The DELPHI collaboration has used Build-Up secondary vertexing, with good results, as an important ingredient in b-tagging [52]. In this case all tracks in the same jet as the seed tracks, are included one-by-one into the vertex fit and the track producing the smallest change in the χ^2 is retained so long as $\Delta\chi^2 < 5$. The procedure is repeated until all tracks satisfying this condition are included in the vertex.

A direct comparison of the Strip-Down and Build-Up approaches to vertexing based on DELPHI data found that the efficiency for finding a b-hadron decay vertex, at a reconstructed decay distance five or more standard deviations from the primary vertex, improved from 29 to 54% when using the Build-Up method. This result should not be surprising since, as we have noted previously, frequently the unique assignment of a track to one vertex over another is not possible within the reconstruction errors. If this is the case, then to select a vertex because it has a small χ^2 could give the wrong answer. The choice between the two methods however is not so clear-cut in practice since the tracks that form the seed vertices are primarily selected by their large displacement from the primary vertex position. This can make the Build-Up methods rather more sensitive to accurate detector resolution modelling in simulation than was the case for the Strip-Down method.

4.4.3.2 Topological Methods

These methods do not place the emphasis on making an initial selection of tracks likely to derive from the secondary vertex, but instead assume that the jets we are interested in will contain tracks that originate from the primary vertex and from (at least) one secondary vertex. The aim is therefore to unravel the decay vertex structure of the hemisphere which best matches the pattern of reconstructed tracks. Although the primary vertex is also reconstructed by such algorithms it is important to note that the best resolution on the interaction point is obtained by the techniques of Sect. 4.4.2 where the tracks are selected from the whole event and so the vertex multiplicities are much higher.

The ALEPH approach from [55] was to determine the difference in $\Delta\chi^2$ between assigning all tracks in the jet to the primary vertex and allowing some to originate from a secondary vertex. The $\Delta\chi^2$ is calculated for candidate secondary vertices located at points on a grid set up in two orthogonal projections containing the jet axis. A vertex is formed using the tracks that are within a three standard deviation contour of the each grid point. The point in space which maximises $\Delta\chi^2$ is determined after interpolating between the grid points around the maximum with a paraboloid surface. An example of this method in action is shown in Fig. 4.13, where a candidate for the decay chain $\bar{B}_s \to e^- D_s^+$ has been found by topologically reconstructing a D_s^+ meson decay into three tracks consistent with $K^+K^-\pi^+$. The resulting $D_s{}^+$ 'track' is then fitted to a common vertex with the electron candidate to form the $\bar{B}_s$ state. This grid-based search method was also used by ALEPH as a primary vertex finder in conjunction with a track pre-selection to remove displaced tracks.

DELPHI [56] developed a similar method which was later elaborated on by OPAL [57]. Starting with a list of reconstructed jets, all tracks in the jet were fitted to a common vertex which was constrained, within the measurement uncertainties, to be the beamspot position. If the resulting χ^2-probability for the fit was larger

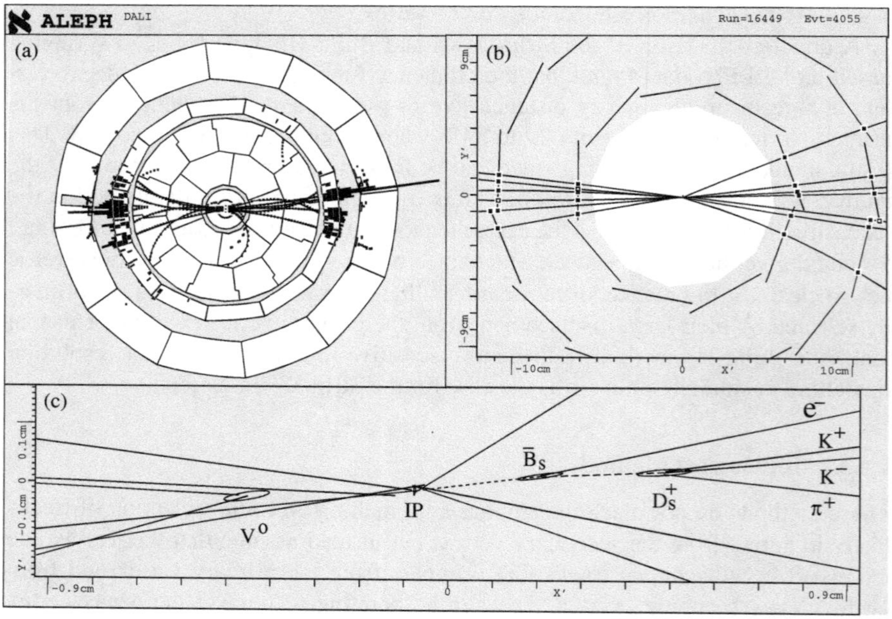

Fig. 4.13 (**a**) A 2-jet $Z^0 \to b\bar{b}$ event reconstructed in the ALEPH detector (**b**) zooming into the radius of the silicon vertex detector (**c**) the event topology viewed well within the radius of the beam pipe, showing the reconstructed positions and error ellipses of the D_s and $\bar{B}_s$ decay points

than 1% the jet was considered to contain no lifetime information and the jet was not considered further. For the jets that remained, the tracks were divided into two groups; one group were fitted to a primary vertex (constrained by the beam spot) and the other group, consisting of at least two tracks, fitted to a secondary vertex. All possible combinations were tried including the case where no tracks were fitted to a primary vertex. For each combination, the combined χ^2 for the primary and secondary fit was formed and that with the best χ^2-probability was selected if:

- the χ^2-probability of the best combination fit exceeded 1%,
- the second best combination fit had χ^2-probability less than 1% and it's χ^2 value was at least 4 units larger than the best fit.

Jets failing these criteria were deemed to be ambiguous and were rejected.

4.4.3.3 An Aside on SLD Topological Vertexing

Although not used at LEP, we mention here the topological vertexing method developed by the SLD collaboration [58] because it is a technique for reconstructing the decay vertex topology in $Z^0 \to b\bar{b}$ events which illustrates what can be achieved if the measurement precision is good enough.

At the Standford Linear Collider, the first measurement point could be very close to the interaction region while the beam spot was very stable in time and very small in size: $(\sigma_x \simeq 2.6\,\mu\text{m}, \sigma_y \simeq 0.8\,\mu\text{m}, \sigma_z \simeq 700\,\mu\text{m})$. These facts allowed the SLD detector to provide 6–7 μm resolution measurements at four points starting at 2.9 cm which was well within the beam pipe radius at LEP. The corresponding impact parameter resolutions were $\sigma(r-\phi) = 11$ μm and $\sigma(z) = 22$ μm, which can be compared to those quoted for the LEP experiments in Table 2.3. More relevant however is impact parameter resolution including the error contribution of the production point. For a typical b-hadron decay product with an impact parameter of 150 μm or so, we have seen that the resolution attained at LEP was in the region of 50 μm. The corresponding number at the SLD was more like 15 μm. With this level of precision it becomes feasible to resolve the decay vertex topology in a hemisphere directly rather than relying on strong track pre-selection cuts which harm the final reconstruction efficiency and/or testing many possible vertex combinations which introduce ambiguities and are time consuming.

The method developed, used the helix parameters of each reconstructed track i to draw a Gaussian probability tube $f_i(r)$ around the 3D track trajectory where the width of the tube is defined by the uncertainty in the track position measured at the point of closest approach to the primary vertex. A vertex probability $V(r)$ was then constructed as a function of the $f_i(r)$ which becomes large when many probability tubes overlap in space. Finally, tracks are associated with maxima in $V(r)$ to form a set of topological vertices as illustrated in Fig. 4.14. After all selection cuts, the method had a total efficiency to resolve a primary and a secondary vertex in a $Z^0 \to b\bar{b}$-event hemisphere of about 50% falling to $\leq$ 5% to find the resulting cascade c-hadron decay vertex. This represents an exemplary performance which the LEP

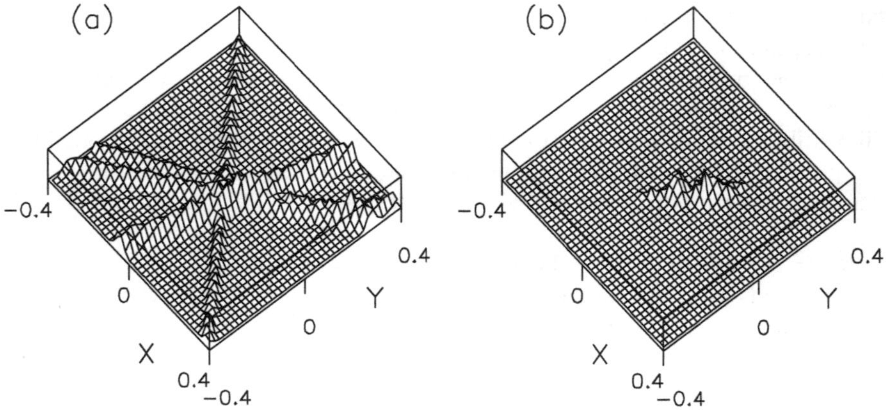

Fig. 4.14 The xy projection of (**a**) $\sum_{i=0}^{N} f_i(r)$ and (**b**) $V(r)$. The trajectories of individual tracks can be seen in (**a**) and the regions where vertices are probable are visible in the distribution of $V(r)$ where the peak at $X = Y = 0$ corresponds to the primary vertex and a secondary peak is evident displaced to the right of the primary by ~ 0.15 cm

experiments could not match because of the resolution issues mentioned earlier. In particular the reconstruction of the cascade vertex, on an event-by-event basis, was impossible at LEP although it was possible to see some level of identification of cascade decay tracks on a statistical basis i.e. averaged over many tracks. This is discussed in Sect. 7.1.4.

It is instructive to mention briefly how SLD turned this vertexing power into a b-tag variable based on the reconstructed mass at the b-hadron decay vertex. The first step was to attach to the topological secondary vertex, further tracks that were consistent with originating from the vertex. As illustrated in Fig. 4.15(a), a vertex axis was drawn joining the centroids of the primary vertex and secondary vertex error ellipses. For each track not in the secondary vertex the 3-D distance of closest approach, T, and the distance along the vertex axis to this point, L, were calculated. Tracks satisfying $T < 1$ mm and $L/D > 0.3$ (where D is the decay length) were then attached to the vertex and the mass of this combination calculated, M_{trks}. The second step involved estimating the amount of missing mass, mainly due to the neutral particles that so far were ignored. The mass of the decaying b-hadron can be written

$$M_{\text{B}} = \sqrt{M_{\text{trks}}^2 + P_{t,\text{trks}}^2 + P_{l,\text{trks}}^2} + \sqrt{M_{\text{miss}}^2 + P_{t,\text{miss}}^2 + P_{l,\text{miss}}^2} \tag{4.11}$$

where M_{miss} is the invariant mass of the set of missed particles. Using the fact that in the *rest frame* of the decaying b-hadron, $P_{t,\text{trks}} = P_{t,\text{miss}} \equiv P_t$ and $P_{l,\text{trks}} = P_{l,\text{miss}}$ and assuming the tracks have pion mass and the neutrals are massless, the *lower* bound for the mass is

$$M_{\text{B}} = \sqrt{M_{\text{trks}}^2 + P_t^2} + |P_t|. \tag{4.12}$$

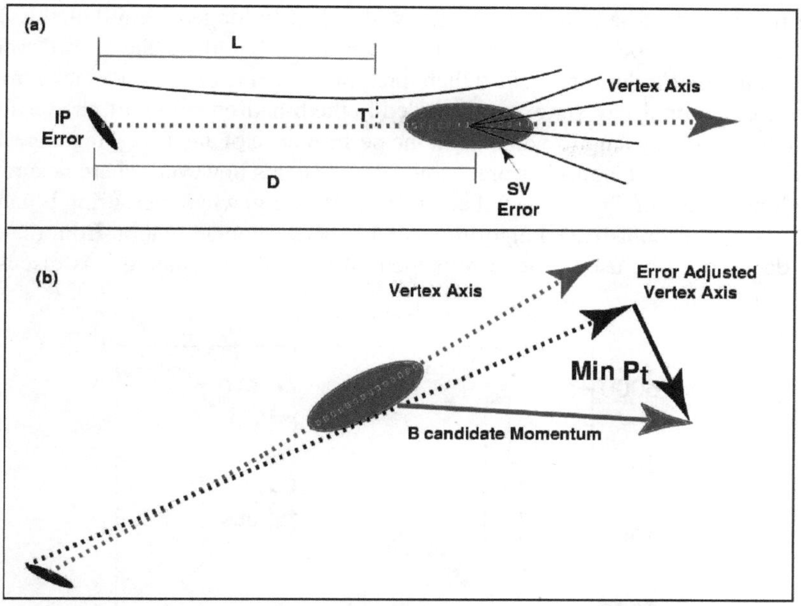

Fig. 4.15 (**a**) The secondary vertex track attachment criteria (**b**) The definition of the vertex minimum missing transverse momentum (P_t)

This P_t corrected mass is suitable for use as a b-tag since the majority of non b-hadron vertices have values that are less than 2.0 GeV/c^2. To reduce the effect of P_t measurements occasionally fluctuating to much larger values than the true P_t, a 'minimum P_t' was calculated by allowing the primary and secondary vertex positions to move anywhere within their error ellipsoids. This is illustrated in Fig. 4.15(b). The minimum P_t is then substituted into 4.12 to give the P_t-corrected mass b-tag which is shown in Fig. 4.16. The performance of the tag is superb: an efficiency for selecting a b-hadron hemisphere of 43.7% and a sample purity of 98.2% is attained where the background consists of 1.5% charm and 0.3% u,d,s hemispheres [59]. This method is clearly dependent on the precision with which the b-hadron flight direction can be reconstructed in order to accurately estimate P_t and P_l. This was possible with the SLD detector because of the excellent primary and secondary vertex reconstruction but attempts to apply similar techniques in the LEP data mainly ended in failure.

4.4.3.4 Closing Remark on Secondary Vertex Reconstruction

With all of the inclusive methods, the final precision on the vertex point depends crucially on selecting only tracks that originated from a common vertex. When fitting for a b-hadron decay vertex the inclusion of tracks from the primary vertex will tend to 'pull' the vertex position to smaller decay lengths, whereas including tracks from the cascade c-hadron decay vertex will pull it to longer decay lengths. This can frequently happen whenever the b-hadron decay length or the B-to-D vertex separation are of order (or smaller) than the typical tracking errors. If tracks from a cascade c-hadron decay vertex are included in the b-hadron vertex fit, the resulting pull to longer decay lengths will benefit the performance of any b-tagging algorithm based on the decay length. For precision measurements however, where accurately modelling details of $B \to D + X$ become important e.g. when measuring b-hadron lifetimes, more sophisticated algorithms that suppress contamination from the cascade decay must be used. The development at LEP of such methods is discussed further in Chap. 7.

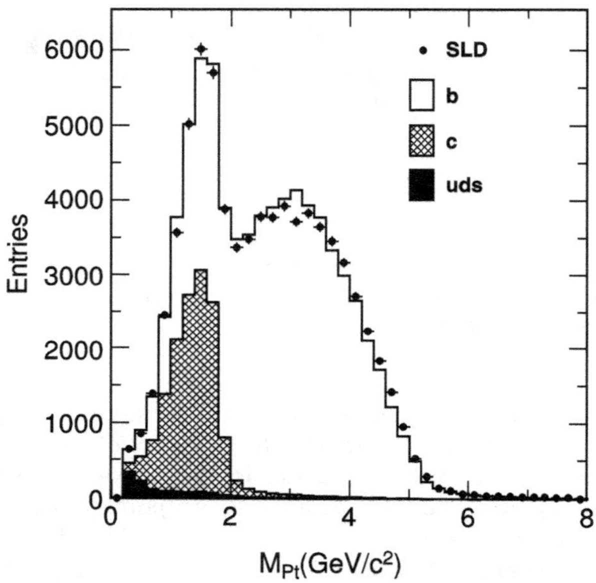

Fig. 4.16 Distribution of the P_t-corrected mass in data (*points*) and simulation (*histogram*). The flavour composition is also indicated: b-events (*open*), c-events (*cross-hatched*) and u,d,s (*dark-shaded*). From [59]

4.4.4 Decay Length Tagging

Experimentally, the decay length is the distance between the reconstructed primary and secondary vertex positions. The variable used for lifetime tagging b-hadrons however is often adjusted for two further effects: The first is to give the decay length a lifetime sign which was defined in Fig 3.1. The second adjustment is to take advantage of any knowledge of the b-hadron flight path when calculating the decay length. As seen in Sect. 4.4.1, the b-hadron direction can be estimated by the jet axis to a precision of about 50 mrad and this can be used to either set the primary

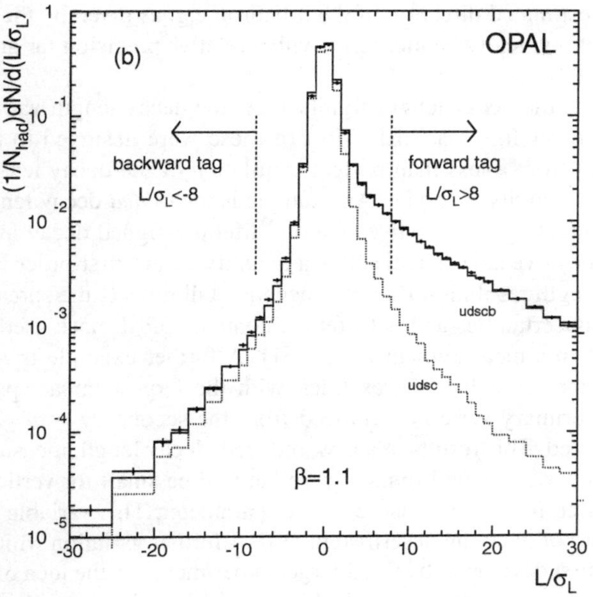

Fig. 4.17 Distribution of decay length significance from OPAL data [61]. The points are the data and the full histogram the simulation prediction. The *dotted histogram* shows the prediction for u,d,s,c-events only

to secondary vertex direction in a fit for the decay length [60], or as a direction constraint already at the secondary vertex fitting stage. Figure 4.17 shows a distribution from OPAL of decay length (scaled by it's calculated uncertainty) where the direction constraint has been imposed on the primary to secondary vertex vector when calculating L. Using the decay length 'significance' L/σ_L, instead of just the decay length, provides a degree of protection against large decay lengths entering an analysis that have been reconstructed with a large associated uncertainty. It can be seen how, with a suitable cut in L/σ_L, a very pure sample of b-hadron decays can be isolated from the u,d,s,c background - note the y-axis is logarithmic!.

In general the tracking resolution of the LEP experiments was superior in the $(R-\phi)$ plane compared to the view in $(R-z)$. Because of this, decay length was often only defined as a 2-D variable and it was not until the introduction of double-sided silicon-strip detectors (see Chap. 2) that full 3-D decay length reconstruction became possible. Even with 3-D vertex detectors it was challenging to match the performance achieved in the $(R-\phi)$ also in the $(R-z)$-plane and it was common practice to use instead the projected 3-D decay length

$$L_{3D} = L_{2D}/\sin\theta. \tag{4.13}$$

Here, θ is the estimated direction of the b-hadron e.g. as given by the jet axis polar angle and would typically be measured with a relative precision far greater than for L_{2D} (see e.g. [62]).

In addition to the decay length significance, the decay length variable has been used in many other forms at LEP. Most of these were designed to reduce uncertainties coming from a less than perfect simulation of the decay length which is a rather complex quantity to model. An example is the *folded* decay length variable – formed by subtracting the negative side of a lifetime-signed decay length distribution from the positive side. The result is a quantity that to first-order is independent of the decay length resolution since the negative tail reflects the spread due only to measurement uncertainties and not lifetime content. OPAL have used folded decay length tagging in a measurement of R_b [61]. A further example is *reduced* decay length. Here, the secondary vertex track with the largest impact parameter with respect to the primary vertex is removed from the secondary vertex track list and the vertex re-fitted. This results in a new, reduced, decay length measurement which for real b-hadron decays will remain large but will be small for vertices formed by one mismeasured track with a large impact parameter. This variable was used, for example, as part of the b-tag described in [53]. A further variation which was used at LEP and was first developed by fixed-target experiments, is the idea of *excess* decay length.[4] This is an attempt to cancel the bias in b-hadron decay length measurements to larger values. The bias comes from two sources: (1) the measured decay length is the convolution of that from the b-hadron with that from the subsequent cascade c-hadron decay and (2) acceptance effects, since experiments tag much more efficiently decays where the b-hadron has traversed a large distance from the production point before decaying. For measurements of b-hadron lifetimes, this bias can be a source of major systematic error if the simulation does not accurately model both the physics and detector effects. For each reconstructed secondary vertex, the method works by finding the minimum decay length that would still result in the secondary vertex being resolved by the analysis selection criteria on such things as secondary vertex fit χ^2. All tracks assigned to the original secondary vertex are assumed to come from the b-hadron decay and are translated back along the b-hadron flight direction (estimated by the jet axis or the primary- to secondary-vertex vector) towards the primary vertex. The excess decay length is then defined to be the translation distance at the point where the arrangement of tracks just fails the secondary vertex selection criteria. This technique was used by both DELPHI [56] and OPAL [57] for b-hadron lifetime measurements and Fig. 4.18 compares the decay length and excess decay length variables from the OPAL analysis. The distributions show that the 'turn-on' shape due to complicated detector acceptance effects is absent in the excess decay length variable which is, to a very good approximation, a negative exponential with the same slope as the b-hadron decay length distribution before secondary vertex selection criteria are applied.

[4] Also sometimes referred to as the reduced decay length in the literature.

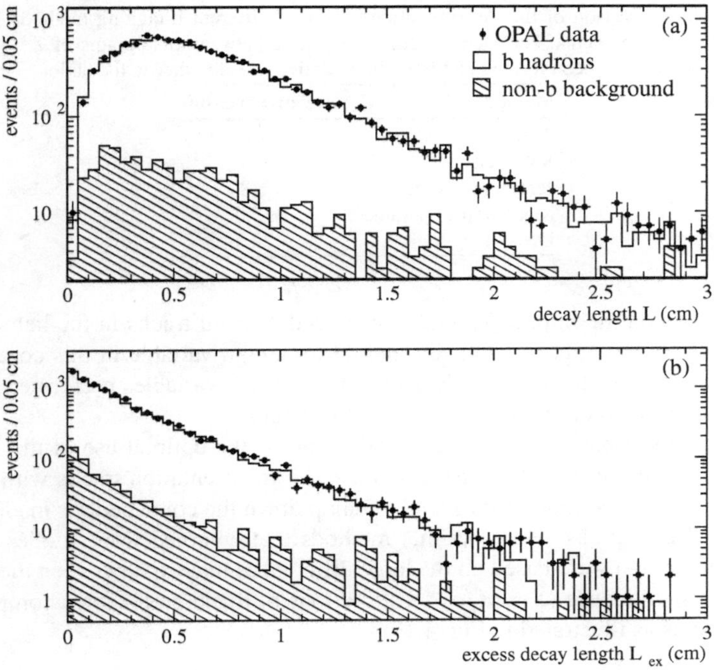

Fig. 4.18 Distributions of (**a**) decay length L and (**b**) excess decay length in data (*points*) and simulation (*histogram*). The expected contribution of b-hadrons and non-b backgrounds are indicated. From [57]

The discriminating power of decay length was in general found to be superior to other methods although the requirement that there be a reconstructed secondary vertex means that the method is often not the most efficient. Furthermore, the decay length was found to be systematically more under control than track impact parameter tags for the following reason: A multitude of reconstruction issues can disrupt any single track impact parameter measurement but reconstructing a secondary vertex point from a bunch of tracks has the additional constraint that all tracks should originate from the same point – so damping down the effects of one badly reconstructed track. This feature meant that the decay length played an important role in the push for precision b-physics measurements.

4.4.5 Impact Parameter Tagging

As a quantity for b-tagging the impact parameter was widely used at LEP. The lifetime-signed impact parameter, defined in Fig. 3.1, is a property of each track and so this method has the advantage that it can be used even if no secondary vertex has been found in the event. Table 4.3 shows the relative discriminating power of

Table 4.3 A comparison of the discriminating power of different b-tagging methods based on DELPHI simulation events. For each variable a χ^2 value between histograms of $Z^0 \rightarrow b\bar{b}$ and background events has been formed and presented relative to each other in the table

Method	Relative measure
Thrust	1.0
Sphericity	4.3
Jet transverse mass	6.8
Mean impact parameter	15.0
Decay length	22.5

a b-tag based on the impact parameter averaged over all tracks in the hemisphere. Although both the impact parameter and decay length variable in this comparison were not optimised, the test clearly shows that lifetime variables performed considerably better than more traditional forms of b-tagging.

Forming the mean impact parameter does not make optimal use of the lifetime information present in a b-hemisphere since the fragmentation tracks with impact parameters scattered around zero tend to damp down the contributions made by the b-hadron decay tracks. More optimal methods that emerged from studies at LEP (discussed below) were based on the large differences that exist between the impact parameter significance ($S = d/\sigma_d$) distributions of tracks in b-events compared to c,u,d,s-events as illustrated in Fig. 4.19.

4.4.5.1 Forward Multiplicity

The forward multiplicity of a hemisphere is determined by counting the number of forward (i.e. $d > 0$) tracks having $d/\sigma_d > S_{\rm min}$, where $S_{\rm min}$ is typically 2.5–3.0. By requiring at least N_f forward tracks in a hemisphere a sample that is essentially 100% b-hadron decays can be formed if N_f is set high enough – see Fig. 4.20. This method, developed by OPAL, was used in early measurements of R_b [64].

4.4.5.2 Impact Parameter Probability Tagging

This method provides a way of constructing a single tagging variable from a group of tracks which could comprise a complete event, one hemisphere or just one jet inside a hemisphere. The idea, first developed by the ALEPH Collaboration [51], is to form from the group of tracks the probability that they are consistent with originating from the primary vertex and the tool used to do it is the significance distribution of Fig. 4.19.

The method is based on a determination of the resolution with which S is reconstructed by measuring the width of the distribution. For this it is important to remove the contribution of tracks originating from displaced secondary vertices since these broaden the distribution for reasons not related to the resolution. This is partially achieved by using only the negative lifetime-signed tail of the distribution. Tracks with negative S values come from one of the following sources: (i) tracks that originate from the primary vertex or close to the primary vertex and pick up a negative

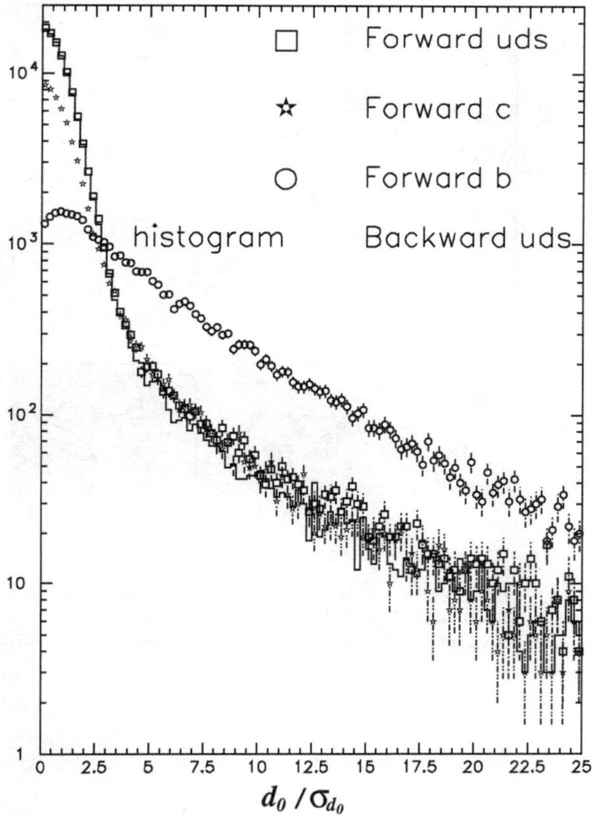

Fig. 4.19 Comparisons of d/σ_d for tracks from b, c and u,d,s-events, from [63]. d is lifetime signed and 'forward' ('backward') imply $d > 0$ ($d < 0$) respectively

sign due to the finite impact parameter resolution; (ii) tracks that originate from a b-hadron secondary decay vertex and pick up a negative sign due to errors in reconstructing the b-hadron direction or primary vertex position; (iii) tracks displaced from the primary vertex, but for reasons not associated with the decay of a long-lived state e.g. tracks from interactions with detector material or reconstruction errors from the tracking algorithm etc. Usually tracks from class (ii) and (iii) can be suppressed by applying any kind of *anti* b-tag to the event but based only on the positive-signed impact parameter tracks in order to leave the negative distribution bias free.

The negative S tail is parameterised with a Gaussian function (or sum of Gaussians) to give the resolution function $f(S)$ which is assumed to be valid for both $S < 0$ and $S > 0$. A track probability function $P(S)$ can now be built which is defined to give the probability for a track, that originates from the primary vertex, to have a impact parameter significance with absolute value S_0 or greater,

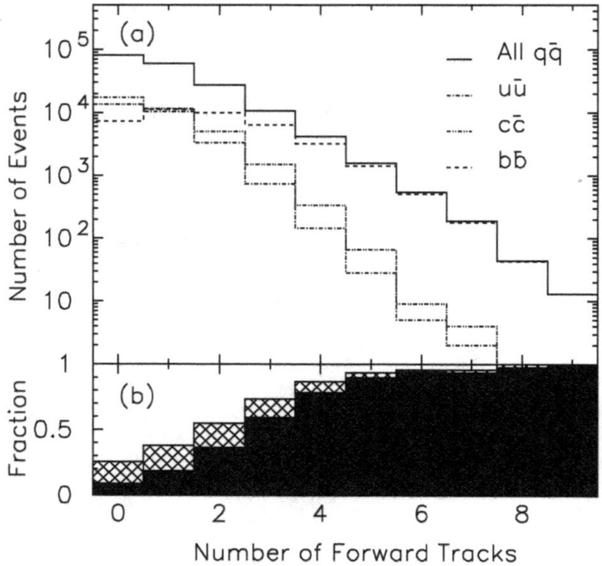

Fig. 4.20 (**a**) The forward multiplicity distribution in Monte Carlo (*solid line*) with the source of the different contributions indicated (**b**) shows the fraction of events in each bin produced by $Z^0 \to b\bar{b}$-events (*black region*) and by $Z^0 \to c\bar{c}$-events (*hatched region*). From [64]

$$P(S_0) = \int_{S=S_0}^{\infty} f(S)\mathrm{d}S \tag{4.14}$$

where $f(S)$ is normalised to an area of one. By construction, for tracks originating from the primary vertex, the distribution of $P(S_0)$ should be flat between zero and one. For tracks from secondary vertices the $P(S_0)$ distribution should peak at low values. Figure 4.21 shows impact parameter probability distributions from simulated events for tracks with positive lifetime-signed impact parameters in $Z^0 \to b\bar{b}$, $Z^0 \to c\bar{c}$ and $Z^0 \to q\bar{q}$ events. The tracks from light quark events are uniformly distributed as expected for a sample originating from the primary vertex, the $Z^0 \to b\bar{b}$ events show a large spike around zero and the $Z^0 \to c\bar{c}$-distribution lies somewhere in between. The small residual spike at zero in the light-quark event distribution is due to contamination from K^0_s and Λ decays.

An important feature of the track probability is that it can be calibrated directly on the data without having to rely on simulations of the impact parameter resolution. In practice this is not such a trivial task however since the resolution is a function of many parameters e.g. momentum, the number of silicon hits on tracks, angle. The resolution is also sensitive to practical issues such as whether some parts of the detector tracking system were inefficient or even turned off for some periods of the data taking. For these reasons, the resolution parameterisation must typically be repeated many times as a function of all quantities that it is found to be sensitive to.

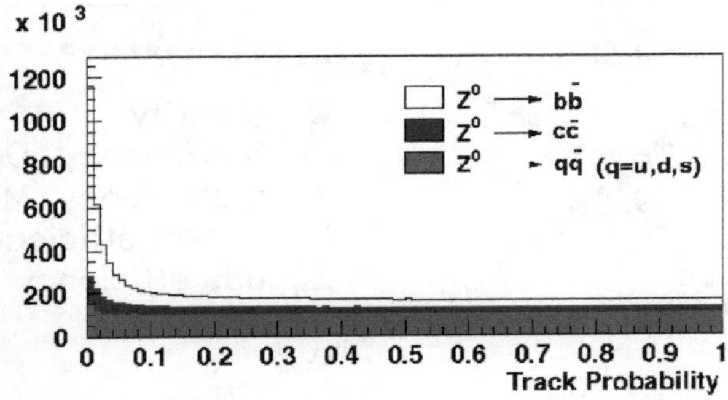

Fig. 4.21 Comparing impact parameter probability distributions in simulated data for $Z^0 \to b\bar{b}$, $Z^0 \to c\bar{c}$ and $Z^0 \to q\bar{q}$-events. From [52]

Similar quantities at the level of a whole group of tracks, e.g. a hemisphere containing n-tracks, can now be defined. If it is assumed that the tracks are independent of each other[5], a reasonable rule with which to combine their probabilities is to take the product, $P(n) = \prod_{i=1}^{n} P_i(S_i)$. The assumption is good enough for most purposes although it is never completely true. A common source of inter-track correlation is if a collection of tracks were fitted together to form a primary vertex. In this case correlation coefficients of order 20% may exist between the tracks.

Having made a measurement of the combined probability $P(n)$, the probability for the group of tracks to have this value $P(n)$ *or greater* can now be calculated as

$$P_G = P(n) \sum_{j=0}^{n-1} \frac{(-\ln P(n))^j}{j!}. \tag{4.15}$$

By forming P_G from tracks with positive impact parameters i.e. from those containing lifetime information, a continuous b-hadron tag is obtained from which different purities and efficiency samples can be selected by cutting on P_G. The formulae can be summed over any group of n tracks so that tags can be formed at the jet (P_J), hemisphere (P_H) or complete event level (P_E). The origin of (4.15) is motivated (a) mathematically as the overall tail probability for a group of probabilities combined by the product rule and (b) experimentally by the observation that no other combination of impact parameter information was found by the LEP experiments to give a better tag of b-quarks.

Figure 4.22 shows the tagging performance obtained by ALEPH from a hemisphere-level tag P_H, where the efficiency quoted is normalised to the number of $Z^0 \to q\bar{q}$ events recorded without any hard selection cuts having been applied. Typically a tagging efficiency of around 30% is possible for a purity of 90% which

[5] Two tracks are independent or uncorrelated if the significance distribution of one is unchanged by making any selection on the other.

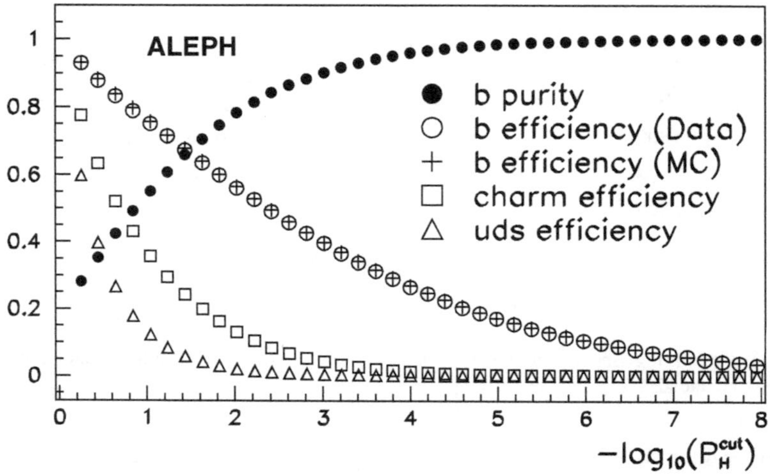

Fig. 4.22 The purity and efficiency performance for tagging $Z^0 \to b\bar{b}$ hemispheres based on the combined probability tag (4.15). From [51]

makes this method very competitive with other techniques. Note that the figure also shows tagging efficiency as measured in the data which is possible using a double-tag technique which will be described in Chap. 6.

Although the resolution functions in data and simulation will differ at some level, the distribution of the tagging variable is flat by construction for any background sample not containing decays of long-lived particles. This means that in principle the simulation can be used to accurately estimate the level of zero-lifetime background in a b-hadron tagged sample. The degree to which this is true can be checked by comparing the tagging efficiency (i.e. the ratio of how many events survive a cut compared to the starting number) between data and simulation using the negative track probability tag which we know is devoid of lifetime content. Backgrounds from sources that contain long-lived states, which means primarily $Z^0 \to c\bar{c}$ events, are harder to control and require a good level of understanding of both charm physics (discussed in Sect. 4.6) and detector resolution effects which were addressed in Chap. 3.

4.4.6 Impact Parameter vs Decay Length

From the point of view of raw b-hadron tagging power, impact parameter based tags and secondary vertex tags can offer rather similar performances. However, when using these tags as part of an analysis, considerations of sensitivity to systematic effects become important and as b-physics developed during the LEP years and measurements became ever more precise, these issues had to be addressed.

The impact parameter of a b-hadron decay product has the nice feature that at high energy, i.e. $\beta \sim 1$, it is independent of the energy of the b-hadron and hence does not rely on a precise knowledge of the fragmentation function. This is in contrast to the decay length and filters through into a marked difference in errors related to the fragmentation function on measurements based on impact parameters compared to decay lengths. In general however, the experience of many LEP analyses e.g. measurements of R_b and of b-hadron lifetimes, has been that impact parameters and related variables tend to be rather more sensitive to details of simulating b-physics and detector effects than methods based on secondary vertex and decay length reconstruction.

An example of this is provided by the time development of the generic b-hadron lifetime τ_b shown in Fig. 4.23. From about 1991 onwards LEP measurements dominated the world average and between 1992 and 1994 the mean had shifted from about 1.3 ps to nearly 1.6 ps which represented a change well outside of the quoted errors. Although the full reasons for this shift were never fully established, the shift occurs at a time when the experiments are including for the first time large quantities of data taken with their precision silicon vertex detectors which in some cases are already in their first upgraded format (see Chap. 2). In addition, the start of resolution tuning procedures as described in Sect. 3.1.5 mean that the simulation of detector resolution was becoming much better understood around this time. The early analyses of lifetimes were based on impact parameter measurements (of leptons and hadrons) and these results were rather drastically affected by the sudden improvements in resolution and modelling of the resolution. Measurements from all experiments drifted upwards during this period. A strong hint that the lifetime was indeed longer than the old impact parameter analyses suggested, came from decay

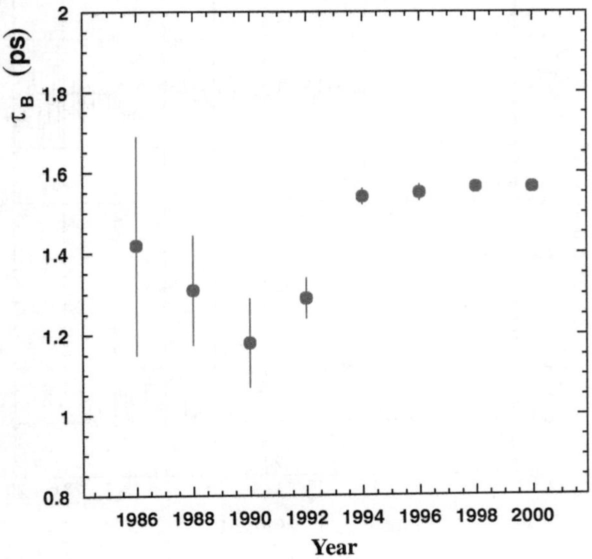

Fig. 4.23 World average mean b-hadron lifetime values as a function of time

length measurements. These began to appear around 1993 and gave systematically higher results that were very much more stable to details of tracking resolution.

4.5 Combined b-Hadron Tagging

When b-hadron tagging variables contain largely orthogonal information the obvious way to try and boost performance is to combine the tags in some way. There were many such examples from LEP, pushed initially by the drive to measure R_b at the one percent level and later by the need to tag possible Higgs decays to b-quarks as efficiently as possible.

A example is provided by the ALEPH lifetime-mass b-tag [44]. It was recognised that cutting on the significance S increased the fraction of b-hadron decay tracks selected up to a maximum and to cut harder just increased the chances of selecting tracks from cascade c-hadron decay. That this is the case is illustrated in Fig. 4.24 from OPAL who found that the fraction plateaued for values of $S \gtrsim 7.0$. To address this, ALEPH developed a tag which was sensitive to the mass difference between c- and b-hadrons which could be used in conjunction with the lifetime tag. The method consisted of combining the four-vectors of tracks in a jet one-by-one in order of decreasing inconsistency with the primary vertex based on their impact parameter probability values $P(S)$ defined in (4.14). This process continues until the invariant mass of the combination exceeds 1.8 Gev/c^2 (the approximate c-hadron mass), at

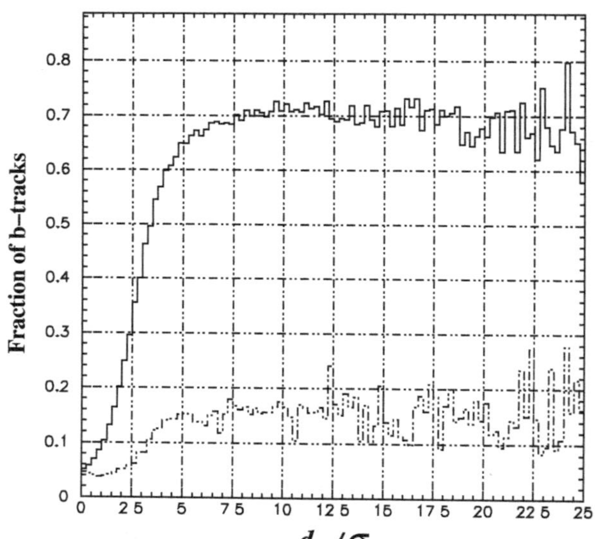

Fig. 4.24 The fraction of all tracks with $d > 0$ (*solid histogram*) and $d < 0$ (*dashed histogram*) that originate from b-hadron decay as a function d/σ_d. From [63]

which point the mass tag for the jet (μ_J) is defined to be the $P(S)$ value of the last track added. For b-jets, μ_J can be rather small since the c-hadron mass can be exceeded only by using tracks from the b-hadron decay, whilst for c-jets, μ_J is larger since tracks from the primary vertex are needed to exceed the same cutoff. At the level of an event hemisphere the mass tag is defined to be $\mu_H \equiv \mu_J(min)$ where $\mu_J(min)$ is the smallest, i.e. the most 'b-like', jet in the hemisphere. ALEPH observed that the mass tag was most effective at rejecting $Z^0 \rightarrow c\bar{c}$ hemispheres which have an unusually large decay length whereas the lifetime tag P_H was most effective at the lower decay lengths. This suggested that some kind of linear combination would make more optimal use of the orthogonal information provided by the two variables and the following combination was found to maximise the b-tag performance: $-\left(0.7\log_{10}\mu_H + 0.3\log_{10}P_H\right)$. The result, at high b-tagging purities, was an increase in tagging efficiency of 50% over that achieved by the lifetime P_H tag alone. A modified version of this tag was also used by OPAL but based on a probability that tracks originate from the primary vertex in place of the the impact parameter probability. This is discussed further in Sect. 7.1.3.

To improve on the b-tagging performance of the impact parameter- based methods, DELPHI made extensive investigations [65, 66], into combining these tags with information gleaned from the reconstruction of secondary vertices. The approach was to form all discriminating variables and the combined b-tag, independently for each jet in an event. The jet clustering was carefully tuned (using the techniques of Sect. 4.4.1) so that, ideally, all the particles associated with the b-quark fragmentation and all those coming from the subsequent b-hadron decay, were included in the jet definition. The DELPHI studies identified the following list of discriminating variables:

(i) The impact parameter *jet probability* P_j^+, as defined in (4.15) where the sums run over all tracks in the jet with positive-sign impact parameters.
(ii) The *effective mass* [6] of particles in the secondary vertex M_S. Figure 4.25(a) shows that the number of vertices in c-jets above the D-meson mass of 1.8 GeV/c^2decreases sharply whereas the mass in b-jets extends up to 5 GeV/c^2.
(iii) The *fraction of the jet energy* X_s^{ch} carried by the charged particles in the secondary vertex, shown in Fig. 4.25(b). For b-hadrons, where almost all the tracks in the secondary vertex are from the b-hadron decay chain, X_s^{ch} is determined by the fragmentation function $f(b \rightarrow \text{hadron})$. The same is true for c-hadrons, but because the fragmentation function $f(c \rightarrow \text{hadron})$ is somewhat softer than for b-fragmentation, the X_s^{ch} distribution is also biased to lower values.
(iv) The *missing transverse momentum* at the secondary vertex P_s^t was first introduced by SLD and was discussed in Sect. 4.4.3.3 as part of the SLD mass tag. It is defined as the magnitude of the resultant momentum vector of all tracks in a

[6] These discriminating variables were calculated assuming charged particles had the rest mass of a pion.

secondary vertex transverse to the estimated b-hadron flight direction (as given by the primary to secondary vertex vector or the jet axis). Because of particles missing from the secondary vertex such as neutrinos from semi-leptonic decay, neutrals or non-reconstructed charged particles, the estimated flight direction and the secondary vertex momentum vector are typically acollinear and P_s^t represents the correction needed to bring them back into line. For all sources of missing energy, the value of P_s^t is expected to be higher for b-quark jets than for c-quark jets because of the high mass of b-hadrons and this is illustrated in Fig. 4.25(c).

(v) The *rapidity* R_s^{tr} of tracks forming the secondary vertex with respect to the jet direction. Figure 4.25(c) shows that the rapidity of particles from b-hadron decay is softer than those from c-hadrons with those from u,d,s events shifted to very low values. The rapidity variable is defined and discussed further in Sect. 7.1.2.

(vi) The *transverse momentum* of a reconstructed, high-energy lepton with respect to the jet axis. The b-tagging power of leptons was discussed in Sect 4.1. They are particularly useful for combined tagging applications since they can be used whether the jet contains a reconstructed secondary vertex or not and the information is independent of lifetime information. In principle the secondary vertex missing transverse momentum variable P_s^t is correlated to lepton production because of the undetected neutrino produced in semi-leptonic decay. However, because semi-leptonic decays are multi-body, the correlation between the lepton and neutrino transverse momenta is relatively weak.

The combination procedure used is worth considering in some detail as a general prescription for combining discriminating variables by a *likelihood ratio*. For a set of variables $\{x_1, \ldots, x_n\}$ that separate a signal source from background, a combined tagging variable y is defined by forming the ratio,

$$y = \frac{f^{\mathrm{bgd}}(x_1, \ldots, x_n)}{f^{\mathrm{sig}}(x_1, \ldots, x_n)}. \tag{4.16}$$

Here, f^{bgd} and f^{sig} are the probability density functions of the discriminating variables for background and signal respectively and the signal is tagged at some purity and efficiency by requiring that y passes some cut value $y < y_{\mathrm{cut}}$. In principle this method gives *optimal* tagging i.e. the best possible background suppression for a given signal efficiency. Note that the likelihood ratio is sometimes also defined as $y = f^{\mathrm{sig}}/f^{\mathrm{bgd}}$ or $y = f^{\mathrm{sig}}/\left(f^{\mathrm{sig}} + f^{\mathrm{bgd}}\right)$ which are functionally related to the definition in (4.16) and will therefore lead to an equivalent performance after a suitable adjustment of the tagging cut.

Unfortunately, handling multi-dimensional probability density functions is in practice difficult for $n > 2$. A way around this is to carefully select variables that are weakly correlated with each other, as was the case for the variables selected by DELPHI for tagging. Under these conditions, it is possible to use (4.16) in the limit of independent variables,

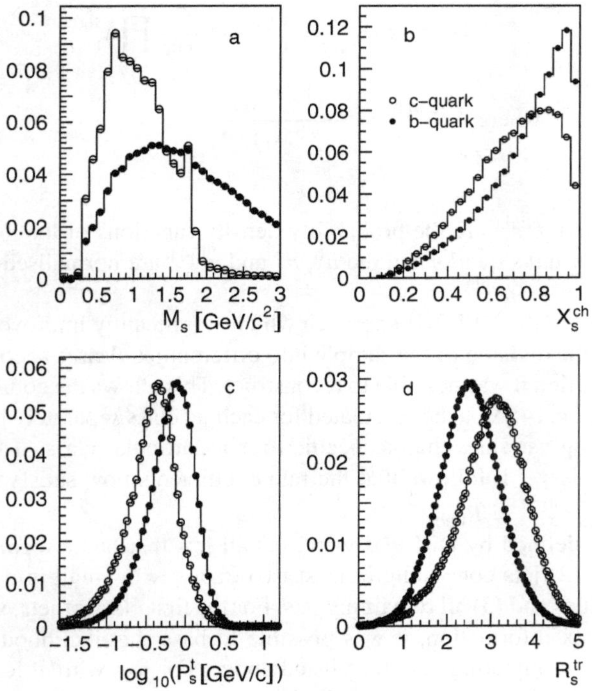

Fig. 4.25 Distributions from simulation, of variables used in DELPHI combined b-hadron tagging. The figures show for b- and c-quark jets separately: (**a**) the secondary vertex mass (**b**) the fraction of the charged jet energy included in the secondary vertex (**c**) the resultant transverse momentum at the secondary vertex (**d**) the rapidity of each track in the secondary vertex. From [52]

$$y = \prod_{i=1}^{n} \frac{f^{\text{bgd}}(x_i)}{f^{\text{sig}}(x_i)} \equiv \prod_{i=1}^{n} y_i \tag{4.17}$$

where $f^{\text{bgd}}(x_i)$ and $f^{\text{sig}}(x_i)$ are probability density functions of each individual variable x_i. Although the combination now described by y is no longer optimal, if the correlations between the variables are small, it is very nearly optimal and has the great advantage that it is easy to form and trivial to extend to more variables. It is often convenient to use y as the transformed variable $y/(y+1)$ so that the full range is [0, 1] and for the case of y being a combined flavour tag (tagging b versus $\bar{\text{b}}$), the transformation $(1-y)/(1+y)$ gives the range $[-1, 1]$.

In the DELPHI approach b-jets were the 'signal' and the 'background' split naturally into two parts: (i) jets originating from c-quarks and (ii) light u,d,s-quark jets. These two components are completely independent and, most importantly, have very different distributions of discriminating variables. This fact means that the background sources should be treated separately in the formalism for the most optimal results an the combined variable was defined as,

$$y = \frac{n^{c}}{n^{b}} \prod_i y_i^{c} + \frac{n^{q}}{n^{light}} \prod_i y_i^{light} \tag{4.18}$$

$$\text{where,} \quad y_i^{c,light} = \frac{f^{c,light}(x_i)}{f^{b}(x_i)}.$$

Here, f^b, f^c and f^{light} are the probability density functions of the x_i for jets generated by b, c or u,d,s quarks and the n^b, n^c and n^{light} are normalised relative rates such that $(n^b + n^c + n^{light}) = 1$.

A final detail to the DELPHI approach which significantly improved the tag performance was the division of the sample into different jet classes according to what tagging information it was possible to reconstruct. This allows the combined tagging variable defined in (4.19) to be calculated for each jet class separately based on a list of discriminating variables that is specific to that particular class. If the combined tag for class α is y_α, it follows that the rate coefficients now satisfy $\sum n_\alpha^b = R_b$, $\sum n_\alpha^c = R_c$, $\sum n_\alpha^{light} = R_{light}$.

The classes defined by DELPHI were: (1) all jets that had a reconstructed secondary vertex, (2) jets containing at least two tracks with small impact parameter probability values and (3) all remaining jets. For the first class of jets, which contain secondary vertex information, it was possible to base the likelihood combination on all of the discriminating variables listed above. For jets with at least two offset tracks, the jet probability, the mass of all offset tracks, their rapidities and any lepton transverse momentum were used. For jets with less than two offset tracks, the mass variable is dropped since there is no longer a reliable core of tracks likely to originate from b-hadron decay and the track rapidities are only used if the track has a positive lifetime-signed impact parameter. It follows that b-jets are tagged with a confidence that increases from the third jet sample to the first and, in this respect, the method of separating the sample into classes itself acts as a additional discriminating variable.

The relative effectiveness of the discriminating variables is illustrated in Fig. 4.26 where the jump in efficiency/purity performance can be seen with the addition of each variable to the jet probability P_j^+.

These results are for tagging b-event hemispheres using the sample of jets that contain secondary vertex information. If the hemisphere contains more than one reconstructed jet, the jet with the lowest value of y is taken to tag the hemisphere. Since b-hadrons are produced in pairs, the presence of a tag from both hemispheres improves significantly the b-tagging of the event as a whole. For many applications, an event b-tag is the starting point to reduce the levels of background before developing an analysis targeted at properties of b-hadrons. The question of how to combine the two hemisphere tags into an event tag is made simpler by the fact that each quark flavour b, c or light, is produced completely independently of any other flavour. For any particular flavour however, one expects correlations to exist between the two hemispheres – this subject is discussed further in Chap. 6 and the simplest approach is to assume these correlations are small and ignore them. In this case, the event b-tag combines the two b-jet tag values as

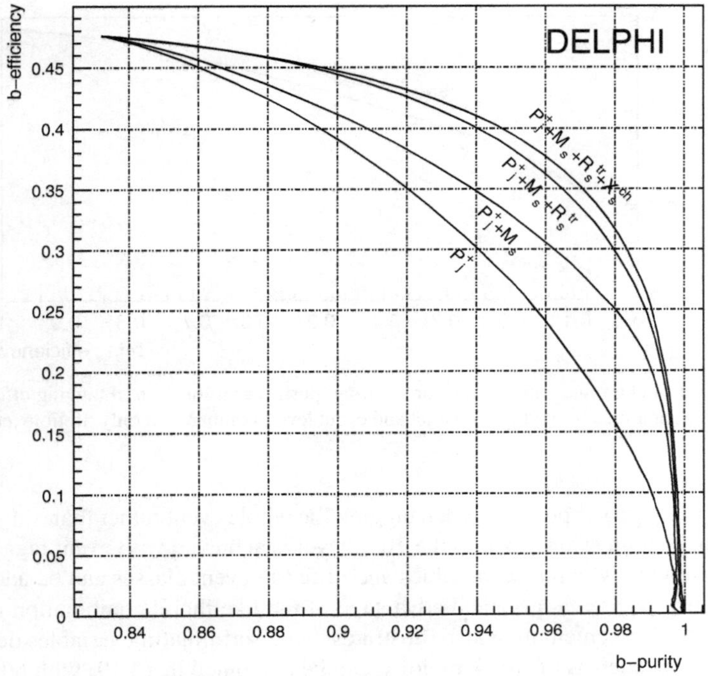

Fig. 4.26 The $Z^0 \to b\bar{b}$-hemisphere tagging purity versus efficiency for the combined tag. From [65]

$$y_{\alpha\beta}^{\text{evt}} = y_\alpha \cdot y_\beta. \tag{4.19}$$

where one jet is from class α and the other from class β.[7] Extending the likelihood formalism to include the effects of hemisphere correlations is also straightforward but results in the more complicated form for the event tag:

$$y_{\alpha\beta}^{\text{evt}} = \frac{R_b}{R_c}\frac{n_\alpha^c n_\beta^c}{n_\alpha^b n_\beta^b}\prod_i y_{\alpha,i}^c \prod_i y_{\beta,i}^c + \frac{R_b}{R_{\text{light}}}\frac{n_\alpha^{\text{light}} n_\beta^{\text{light}}}{n_\alpha^b n_\beta^b}\prod_i y_{\text{alpha},i}^{\text{light}} \prod_i y_{\beta,i}^{\text{light}}. \tag{4.20}$$

The performance of the event-level tag is compared in Fig. 4.27 to the jet-level combined tag and a jet tag based only on the jet probability. The plot shows the huge gains that combined tagging brings over lifetime tagging alone and, as expected, the

[7] For cases where the event has more than two jets, the smallest-two jet tag values can be used.

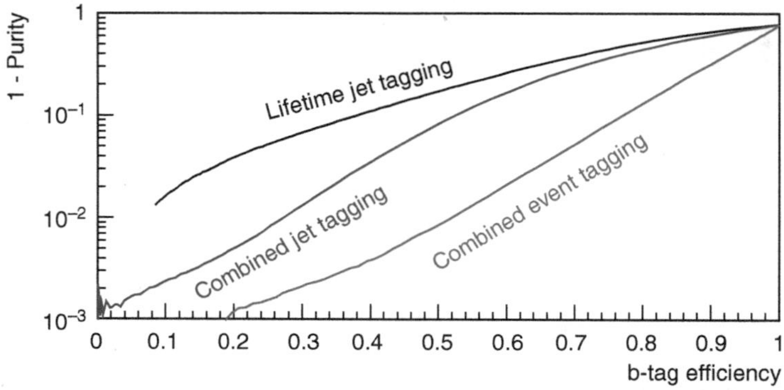

Fig. 4.27 The background suppression (or 1− b-tag purity) as a function of tagging efficiency in Z^0decays for combined b-tagging (at the jet and event level) compared to only lifetime jet tagging. From [52]

gain in tagging power possible when tagging the whole event rather than single jets – background suppression down to the 10^{-3} level is achieved with event tagging.

The ease with which new variables and different event classes can be added into the formalism is a great practical advantage of the likelihood combination method. The DELPHI implementation also illustrates that discriminating variables defined at the track level, such as the track rapidity, can be combined in (4.19) with other variables defined only at the jet level. There are however two limitations of the method. The first problem has already been discussed – namely that the combination assumes input variables that are uncorrelated with each other. The second limitation is that the method can be heavily dependent on accurate modelling by the simulation. For b-tagging, the probability density functions for signal and background must almost always come from simulation.[8]

Because of the issue of correlations, ultimately the best performing b-tags from LEP were based on combinations made by artificial neural networks. In principle, neural networks can be trained to 'learn' the often very complicated correlations between the discriminating variables in order to find the optimal combination. In contrast to the likelihood combination, discriminating variables that are correlated with each other but which none-the-less carry some degree of orthogonal tagging information will in principle still improve the overall performance of the network.

An early application of neural networks for b-tagging was in the combination of event shape and lepton variables [67–69]. Later, networks appeared that used discriminating variables based on a wider range of physics properties and, in particular, the lifetime tags. Incredible tagging power was possible from such approaches, an example of which is the tag developed by L3 [70]. The method combines 14 vari-

---

[8] For some applications it may be possible to derive the probability density functions directly from the data e.g. when using a likelihood method to identify electrons, a clean sample of electrons can be obtained by reconstructing photon conversions.

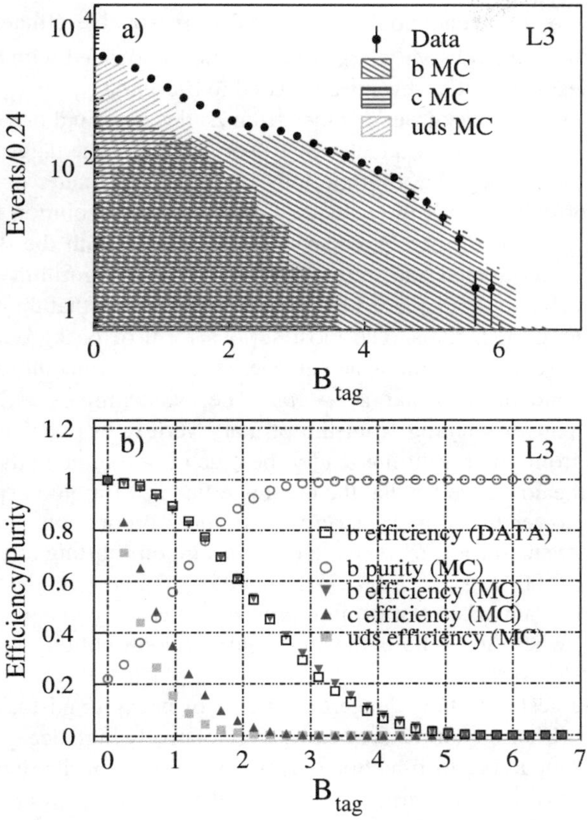

Fig. 4.28 (**a**) The spectrum of the combined b-tag for $Z^0 \rightarrow q\bar{q}$-events at 91 Gev and (**b**) the purity versus efficiency performance for tagging $Z^0 \rightarrow b\bar{b}$-events

ables in total including: lifetime tags, variables reconstructed from secondary vertices such as invariant mass and multiplicity, event shape variables such as boosted sphericity and the momenta of lepton candidates. The network output, given in Fig. 4.28(a), shows that there is a large region of the parameter space for which $Z^0 \rightarrow b\bar{b}$ events can be tagged with essentially no background. The resulting tag performance is shown in Fig. 4.28(b) and indicates that a 100% pure sample of $Z^0 \rightarrow b\bar{b}$ events can be isolated with an efficiency of around 20%. Similar tags with equally impressive performances were developed by ALEPH [71] and OPAL [72].

4.6 Background Issues

Precise and reliable inclusive b-physics measurements were only possible at LEP after b-tagging techniques had reached the stage where very pure samples (purities

$\gtrsim 90\%$) of $Z^0 \to b\bar{b}$ events could be isolated at reasonable efficiency. A major reason for this were the relatively large uncertainties associated with the modelling of physics processes in u,d,s,c-events compared to b-events.

The least dangerous background comes from Z^0 decays to u,d or s-quarks which can be routinely suppressed by a factor of 100 or so by lifetime tagging techniques. Moreover, use of the negative lifetime tails in impact parameter or decay length distributions normally allows the detector efficiency and resolution to be reliably extracted. Tracks from K^0_s and Λ decays or interactions with the detector material are easily reconstructed and removed by specialised algorithms and any that remain are typically so far displaced from the primary vertex that they are excluded from the b-tagging algorithms. The most serious form of background from light quark events that remains, occurs when an energetic gluon splits into a heavy quark pair: $g \to c\bar{c}$ and in particular $g \to b\bar{b}$. The event mimics a Z^0 decay with a heavy quark pair originating from the primary vertex. In practice however the heavy hadrons from gluon splitting tend to be much less energetic than those from Z^0 decay with the consequence that the tagging efficiency for gluon splitting events is significantly lower than for primary c- or b-quark production. In addition, the absolute rates are small for a Z^0 decay to contain a gluon splitting to a heavy quark: $g_{c\bar{c}} = 0.0296 \pm 0.0038$, $g_{b\bar{b}} = 0.00254 \pm 0.00051$ [73]. Nevertheless these effects do become significant when performing precision measurements e.g. in the measurement of R_b where $g \to b\bar{b}$ is the single largest contribution to the systematic error on the world average.

By far the largest and most dangerous source of background for b-physics are $Z^0 \to c\bar{c}$ events. Charm hadrons are heavy, with relatively high decay multiplicities and lifetimes comparable to b-hadrons. This makes the contribution of charm to b-tagged samples difficult to unravel – a process that almost always requires the use of a Monte Carlo simulation. However, accurate modelling of $Z^0 \to c\bar{c}$ events is hindered by the current status of experimental knowledge in the charm sector. The following, non-exhaustive list, summarises the problem areas roughly in the order of seriousness to b-physics analyses:

- Lifetime differences of more than a factor four mean that tagging efficiencies vary significantly for the different types of c-hadrons.
- A complete database of c-hadron decay branching ratios does not exist. The best measured channels are those most easily reconstructed experimentally. The current precision of the branching ratios for these ‘reference channels’ can be seen in Table 1.2 and these feed directly into the systematic uncertainty of many b-physics analyses.

  In order to estimate from simulation the tagging efficiency in $Z^0 \to c\bar{c}$ events, knowledge of inclusive branching ratios are important but mostly consist only of relatively old measurements taken at the Ψ''. Branching ratios for the decay $D \to K^0X$ are of particular importance since in this case the K^0 carries quite a lot of the energy and invariant mass so that the hemisphere containing the decay is often not tagged. These branching ratios are known however to no better than 10%.

Knowledge concerning the properties of the excited states D^* and D^{**} is also limited. Since the D^{*+} can decay into $D^0\pi^+$ and $D^+ (\pi^0 \text{ or } \gamma)$ but the (lighter) D^{*0} only into $D^0 (\pi^0\gamma)$, the ratio of D^+/D^0 becomes difficult to predict. The production of D^{**} states from B-decays which subsequently decay to a D^* is a major source of systematic uncertainty in exclusive $|V_{cb}|$ measurements.

- The D production fractions i.e. $f_{D_i} = BR(\bar{c} \to \bar{D}_i) = BR(c \to D_i)$ where $D_i = D^+, D^0, D_s^+$ or charm baryon, were measured at LEP simultaneously with R_c in charm-counting analyses by ALEPH [19], DELPHI [12] and OPAL [74, 75]. In practice the analyses extract products of the form, $R_c \times f_{D_i} \times BR\,(D_i \to X)$ where $BR\,(D_i \to X)$ is the value of the relevant reference channel branching ratio. Hence the errors on the reference branching ratios feed directly into errors on the f_{D_i}. After correcting for the decay channel branching ratios, the measurements enter the electroweak heavy flavour fit described in Sect. 1.4 where R_c and the fractions f_{D_i} are fitted parameters. One complication to this scheme are the strange-charm baryons for which no published rate measurements exist. Their contribution is estimated to be 15±5% of the Λ_c^+ rate by extrapolating from measurements in the light quark sector and assuming Ξ_c^0 contributes the same as Ξ_c^-. These analyses also extract the product, $R_c \times f_{c \to D^{*+}} \times BR\,(D^{*+} \to \pi^+ D^0) \times BR\,(D^0 \to K^-\pi^+)$, using the techniques described in Sect. 4.3 and where D^0 is reconstructed in the reference channel $D^0 \to K^-\pi^+$. The product also enters the heavy flavour fit and the result must be corrected for $BR\,(D^{*+} \to \pi^+ D^0)$ before $f_{c \to D^{*+}}$ can be extracted.
- Knowledge of D meson decay multiplicities come from a single experiment, MARK-III, and are now rather old [76]. The efficiency of tagging algorithms which search for tracks displaced from the primary vertex are, of course, rather sensitive to whether e.g. a D^+ decayed in a 1,3 or 5-prong topology.
- The fraction of the beam energy attained by c-hadrons in fragmentation is now accurately constrained by LEP analyses using lepton tags or inclusive reconstruction of D^0/D^+-mesons[74] and by analyses with full D^{*+} reconstruction as mentioned in Sect. 4.3. The value used by the EWWG is, $\langle x_c \rangle = 0.484 \pm 0.008$ [73].

Many of these uncertainties on charm parameters have also direct consequences for modelling $Z^0 \to b\bar{b}$ decays through cascade decays i.e. the process $B \to DX$ and the subsequent D-decay.

4.6.1 The 'R_b-crisis'

We end this chapter with an illustration of how vital high performance $Z^0 \to b\bar{b}$ tagging was to precision b-physics at LEP – the 'R_b-crisis' of the mid-1990s. The introduction of the double-hemisphere method to extract tagging efficiencies direct from data (see Chap. 6) meant that R_b measurements could approach the level of precision needed to start probing new physics contributions. Discrepancies with Standard Model predictions for R_b soon began to develop until in 1995, the difference between the world average and theory stood at 3.7σ. The subsequent development of R_b with time is illustrated in Fig. 4.29 which shows that it took a few years for the value to finally stabilise at a level consistent with the Standard Model prediction. This coincided with the introduction of 'multi-tag' methods (see Sect. 6.3) which provided a much more effective rejection of charm events than was previously possible (at least some of the effect was also due to analyses with a lower level of 'hemisphere correlation' which is discussed further in Chap. 6). The danger of claiming precision measurements without having a full control of the backgrounds is highlighted by the profusion of theoretical papers that appeared around 1996 claiming to 'explain' the discrepancy with models ranging from SUSY based solutions (e.g. [77]) to theories containing new U(1) symmetries with leptophobic gauge bosons! [78]. A further illustration of the battle LEP b-physics had with charm backgrounds has already been seen in Fig. 4.23. Without doubt, a contribution to the step upwards in τ_b that occurred between about 1993 and 1994 was the much improved rejection of the charm background that became possible with fully working and well understood silicon vertex detectors in the LEP experiments. In hindsight it is rather clear that the errors on early measurements of R_b and τ_b, which all used semi-leptonic B-decays, were underestimated. The significance therefore of the effects in Fig. 4.29 were probably not as large as reported at the time. These examples do

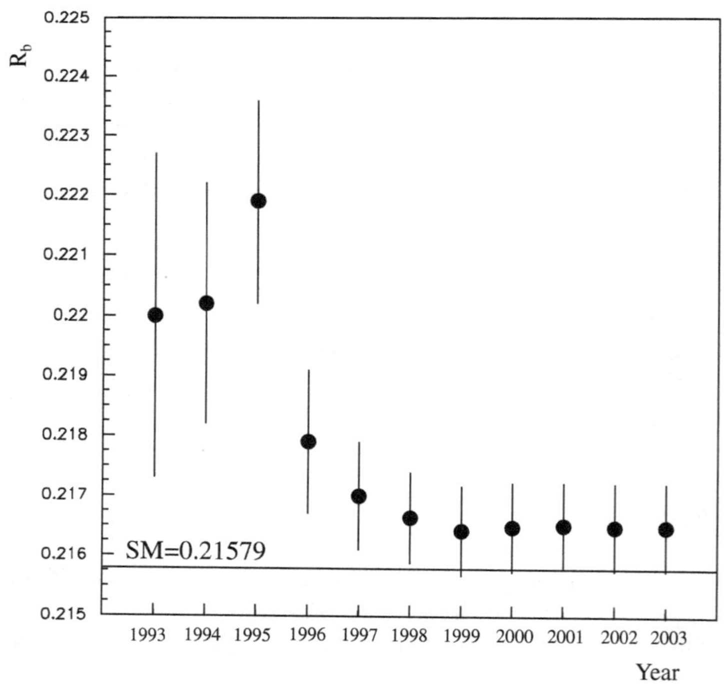

Fig. 4.29 World average values for R_b and as a function of time. The Standard Model prediction is also indicated

however illustrate the crucial role that high performance b-tagging played in the development of b-physics analyses at LEP.

4.7 Summary

Inclusive b-physics demands the reconstruction of an efficient 'tag' of the underlying properties of $Z^0 \to b\bar{b}$ events. In the LEP environment, traditional methods of tagging $Z^0 \to b\bar{b}$ events such as high p_t leptons, events shapes or D^* mesons were found to be much less powerful than lifetime based tags. Impact parameter and secondary vertex techniques were both widely used and found to give similar tagging performance. In the later stages of LEP where techniques had developed to the stage where precision b-physics measurements were possible, secondary vertex techniques were often found to be rather less sensitive to details of modelling the track reconstruction in the detectors.

Inclusive methods also often demand use of Monte Carlo and detailed detector simulations to model the various contributions to the inclusive-variable distributions. The rejection of charm background was proven to be essential in precision measurements such as R_b and τ_b and took a number of years before techniques had reached the stage where background systematics were fully under control.

References

1. N. Isgur, D. Scora, B. Grinstein: Phys. Rev. D **39**, 799 (1989)
2. G.J. Barker: A measurement of the mean *b*-hadron lifetime using the central drift chambers of the OPAL experiment at LEP. PhD Thesis RALT-175, University of London (1992)
3. R. Marshall: Z. Phys. C **26**, 291 (1984)
4. TASSO Collab., W. Baunschweig et al.: Z. Phys. C **44**, 1 (1989)
5. DELPHI Collab., P. Abreu et al.: Phys. Lett. B **281**, 383 (1992)
6. MARKII Collab., O.J. Yelton et al.: Phys. Rev. Lett. **49**, 430 (1982)
7. ALEPH Collab., D. Buskulic et al.: Phys. Lett. B **352**, 479 (1995)
8. ALEPH Collab., R. Barate et al.: Phys. Lett. B **434**, 415 (1998)
9. DELPHI Collab., P. Abreu et al.: Eur. Phys. J. C **10**, 219 (1999)
10. OPAL Collab., G. Alexander et al.: Z. Phys. C **73**, 379 (1996)
11. ALEPH Collab., R. Barate et al.: Eur. Phys. J. C **4**, 557 (1998)
12. DELPHI Collab., P. Abreu et al.: Eur. Phys. J. C **12**, 225 (2000)
13. OPAL Collab., K. Ackerstaff et al.: Eur. Phys. J. C **1**, 439 (1998)
14. ALEPH Collab., D. Buskulic et al.: Z. Phys. C **62**, 1 (1994)
15. DELPHI Collab., P. Abreu et al.: Z. Phys. C **59** 533 (1993)
16. Erratum: Z. Phys. C **65**, 709 (1995)
17. OPAL Collab., R. Akers et al.: Z. Phys. C **67**, 27 (1995)
18. OPAL Collab., G. Abbiendi et al.: Eur. Phys. J. C **13**, 1 (2000)
19. ALEPH Collab., R. Barate et al.: Eur. Phys. J. C **16**, 597 (2000)
20. OPAL Collab., K. Akerstaff et al.: Eur. Phys. J. C **5**, 1 (1998)
21. DELPHI Collab., P. Abreu et al.: Z. Phys. C **71**, 539 (1996)
22. OPAL Collab., G. Abbiendi et al.: Phys. Lett. B **493**, 266 (2000)
23. DELPHI Collab., P. Abreu et al.: Z. Phys. C **74**, 19 (1997)
24. DELPHI Collab., P. Abreu et al.: Z. Phys. C **76**, 579 (1997)

25. K. Hagiwara et al.: Phys. Rev. D **66**, 010001 (2002) and 2003 off-year partial update for the 2004 edition available on the PDG WWW pages (URL: http://pdg.lbl.gov/)
26. M.A. Shifman, M.B. Voloshin: Sov. J. Nucl. Phys. **47**, 511 (1988)
27. N. Isgur, M. Wise: Phys. Lett. B **232**, 113 (1989)
28. N. Isgur, M. Wise: Phys. Lett. B **237**, 527 (1990)
29. A.F. Falk, H. Georgi, B. Grinstein, M.B. Wise: Nucl. Phys. B **343**, (1990)
30. M. Neubert: Phys. Lett. B **264**, 455 (1991)
31. M. Neubert: Phys. Lett. B **338**, 84 (1994)
32. OPAL Collab., G. Abbiendi et al.: Phys. Lett. B **482**, 15 (2000)
33. M.E. Luke: Phys. Lett. B **252**, 447 (1990)
34. M. Neubert: Introduction to B Physics. In:*Lectures presented at the Trieste Summer School in Particle Phsyics (Part II)*, hep-ph/0001334 (2000)
35. ALEPH Collab., D. Buskulic et al.: Phys. Lett. B **395**, 373 (1997)
36. DELPHI Collab., P. Abreu et al.: Phys. Lett. B**510**, 55 (2001)
37. DELPHI Collab., J. Abdallah et al.: Eur. Phys. J. C **33** 213 (2004)
38. I. Caprini, L. Lellouch, M. Neubert: Nucl. Phys. B **530**, 153 (1998)
39. CLEO Collab., J. Bartelt et al.: Phys. Rev. Lett. **82**, 3746 (1999)
40. BELLE Collab.,K Abe etal: Phys. Lett. B**526**, 258 (2002)
41. In: *The CKM Matrix and the Unitarity Triangle. Workshop, CERN, Geneva, Switzerland, 13-16 Feb 2002*, M. Battaglia, A.J. Buras, P. Gambino, A. Stochhi (eds.) (CERN-2003-002, FERMILAB-CONF-02-422, June 2003) p. 288
42. JADE Collab., W. Bartel et al.: Z. Phys. C **33**, 23 (1986)
43. S. Moretti, Lönnblad, T. Sjöstrand: New and Old Jet Clustering Algorithms for Electron-Positron Events. hep-ph/9804296 (1998)
44. ALEPH Collab., R. Barate et al.: Phys. Lett. B **401**, 150 (1997)
45. PLUTO Collab., Ch. Berger et al.: Phys. Lett. B**97**, 459 (1980)
46. R. Kowalewski: Study of jet finding for b physics in OPAL. OPAL Technical Note TN180 (1993)
47. ALEPH Collab., A. Heister et al.: Eur. Phys. J. C **24**, 177–191 (2002)
48. T.Sjöstrand: Comp. Phys. Commun. **28**, 227 (1983)
49. DELPHI Collab., P. Abreu et al.: Phys. Lett. B **345**, 598 (1995)
50. Z. Albrecht, T. Allmendinger, G. Barker, M. Feindt, C. Haag, M. Moch: BSAURUS-A Package For Inclusive B-Reconstruction in DELPHI., hep-ex/0102001 (2001)
51. ALEPH Collab., D. Buskulic et al.: Phys. Lett. B **313**, 535 (1993)
52. DELPHI Collab., J. Abdallah et al.: Eur. Phys. J. C**32**, 185 (2004)
53. OPAL Collab., G. Abbiendi et al.: Eur. Phys. J. C **8**, 217 (1999)
54. ALEPH Collab., D. Buskulic et al.: Phys. Lett. B **388**, 648 (1996)
55. ALEPH Collab., D. Buskulic et al.: Phys. Lett. B **322**, 441 (1994)
56. DELPHI Collab., W. Adam et al.: Z. Phys. C **68**, 363 (1995)
57. OPAL Collab., G. Abbiendi et al.: Eur. Phys. J. C **12** 609 (2000)
58. D.J. Jackson: Nucl. Instr. Meth. A **388**, 247 (1997)
59. K. Abe et al.: Phys. Rev. D **65** 092006 (2002)
60. R. Ong, K. Riles: The Decay Length Method. MARKII/SLC Note 166 (1986)
61. OPAL Collab., K. Ackerstaff et al.: Z. Phys. C **74**, 1 (1997)
62. OPAL Collab., K. Ackerstaff et al.: Zeit. Phys. C **73**, 397 (1997)
63. R. Kowalewski: Study of lifetime tagging of b hadrons in OPAL. OPAL Technical Note TN181 (1993)
64. OPAL Collab., P.D. Acton et al.: Zeit. fur Physik C **60**, 579 (1993)
65. DELPHI Collab., P. Abreu et al.: Eur. Phys. J. C **10**, 415–442 (1999)
66. G. Borisov: Nucl. Instr. Meth. A **417**, 384 (1998)
67. ALEPH Collab., D. Buskulic et al.: Phys. Lett. B **313**, 549 (1993)
68. DELPHI Collab., P. Abreu et al.: Phys. Lett. B **295**, 383 (1992)
69. L3 Collab., O. Adriani et al.: Phys. Lett. B **307**, 237 (1993)

70. L3 Collab., M. Acciarri et al.: Phys. Lett. B **411**, 373 (1997)
71. ALEPH Collab., R. Barate et al.: Phys. Lett. B **412**, 173 (1997)
72. OPAL Collab., G. Abbiendi etal: Eur. Phys. J. C **26**, 479 (2003)
73. LEP/SLD Heavy Flavour Working Group: Final input parameters for the LEP/SLD heavy flavour analyses. LEPHF/2001-01
74. OPAL Collab., G. Alexander et al.: Z. Phys. C **72**, 1 (1996)
75. OPAL Collab., K. Ackerstaff et al.: Eur. Phys. J. C **1**, 439 (1998)
76. MARK-III Collab., D. Coffman et al.: Phys. Lett. B **263**, 135 (1991)
77. D. Choudhury , D.P. Roy: Phys. Rev. D **54**, 6797 (1996)
78. K.S. Babu, C.F. Kolda, J. March-Russell: Phys. Rev. D **54**, 4635 (1996)

Chapter 5
Tagging b-quark Charge

For some b-physics topics it is not enough to just tag the presence of b-quarks – it is necessary to also know whether the quark or anti-quark state is in the data. Examples include measuring the forward-backward asymmetry A^{b}_{FB} which must distinguish b- from $\bar{b}$ at their point of production or measurements of neutral B-meson oscillations for which knowledge of the b-quark flavour is required at both the production and the decay time. Clearly, in order to achieve this a method that can somehow tag the quark charge must be employed. Note that this procedure is often (rather confusingly) termed 'flavour tagging'.

The physics of b-production and decay throws up a number of opportunities to indirectly try and piece together what the charge of the underlying quark was in the event hemisphere. Experimentally however we are limited by those particles which can be readily reconstructed and so full use must be made of: (i) particle ID techniques (in order to find leptons and kaons for example) and (ii) the lifetime tagging techniques described in Chap. 4 to tag whether these particles originate from the primary vertex or from some displaced decay vertex.

5.1 Production Flavour

Figure 5.1 illustrates some examples of associated particle production during the fragmentation of b-quarks, that are useful in determining the charge of the parent quark. For example, if a $\bar{b}$ quark combines with a u-quark to form a B^+ meson, the $\bar{u}$ -partner may combine with a d-quark to form a π^- (similarly if a b-quark hadronises to form a B^- meson, the associated pion would be a π^+). These correlations get enforced by the production of the P-wave B^{**} states through their decays to the pseudoscalar ground states: in our example, a B^+ meson produced in association with a π^- could also originate from the decay $B^{0**} \to B^{(*)+}\pi^-$, and in both cases the correlation would signal the presence of an underlying b-quark. On the down side however, the tag has now become rather complicated since the B^{**} decay and the fragmentation process have both produced a pion but of opposite charges. Usually no attempt is made to try and separate the fragmentation tracks from the B^{**} decay products and so modelling the branching fraction of b-quarks to

G.J. Barker, *b-Quark Physics with the LEP Collider*, STMP 236, 111–119,
DOI 10.1007/978-3-642-05279-8_5, © Springer-Verlag Berlin Heidelberg 2010

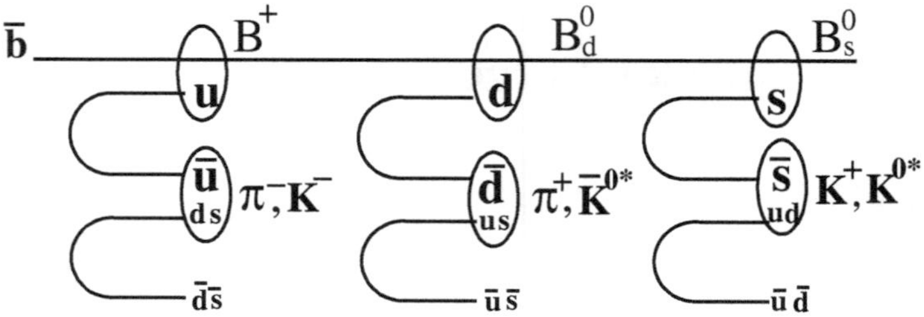

Fig. 5.1 Schematic picture of $\bar{b}$ fragmentation illustrating the charge correlation of the parent $\bar{b}$ quark to the various leading fragmentation particles that can be appear. The charge conjugate processes occur with all charge signs reversed

B** states, which could be as high as 30% (see Sect. 1.3.6), is a potential systematic uncertainty for flavour tagging based analyses. Figure 5.1 also shows that charge correlations can be expected based on kaons.

In all cases, identifying these production flavour tracks is not easy but is clearly a task that can be made easier if reliable particle identification is available. In addition these tracks originate from the primary interaction vertex and have trajectories that follow quite closely the path of the parent b-quark. This means that quality tracking, vertex finding and jet reconstruction are also valuable tools to help select the tracks of interest from the large number of background tracks. We now discuss two reconstructed quantities that were used at LEP for production flavour tagging.

5.1.1 *Jet Charge*

It has long been known that the sum of particle charges associated with a quark jet, averaged over a sample of jets, has a mean value that is correlated with the charge of the quark that initiated the jet [1]. The fragmentation process can be regarded as a hierarchical chain with the hadron containing the original quark occupying the highest position or rank (see e.g. Fig. 5.1). Models of the fragmentation process predict, and subsequent experimental measurements have shown, that the particles produced highest in the hierarchy 'remember' the charge of the parent quark but that this correlation grows progressively weaker as one considers particles produced from lower down the order. Since particles high in the hierarchy are more likely to be produced with large momentum and rapidity with respect to the parent quark direction, a sensible strategy for reconstructing the quark charge would be to weight in some way the charges of the jet tracks based on their momentum and/or rapidity. Such a variable is the jet charge, which has been used extensively by the LEP experiments (and before LEP) and is defined as

$$Q_{\text{jet}} = \frac{\sum_i q_i \cdot |p_i \cdot T|^{\kappa}}{\sum_i |p_i \cdot T|^{\kappa}} \tag{5.1}$$

for q_i the measured track charge, T some estimate of the parent b-quark direction (e.g. the event thrust axis or a jet axis) and the sum usually runs over all tracks thought to be associated with the quark jet. The parameter κ typically takes on different values according to the type of quark jet one wishes to flavour tag and the experimental conditions. The consensus from various studies at LEP was that a value of κ in the range [0.3–0.8] gave the optimal charge separation between b and $\bar{\text{b}}$ for unbiased jets but that a lower value was more suited if a jet charge based mainly on the fragmentation tracks was required i.e. after removing the (high momentum) b-hadron decay products from the jet. Jet charges based on different κ-values are correlated but choosing a range of values was often found to bring some extra information, albeit small, which was useful e.g. as part of the input to a flavour tagging neural network (see below). Jet charge works well for all species of b-hadrons.

5.1.2 Weighted Primary Vertex Charge

Vertex charge is a generic term used to describe a variable sensitive to the charge at a secondary vertex. It is based on a track-by-track measure of the probability for a track to originate from a reconstructed secondary vertex in the event hemisphere (P_{PS}). Examples of such probabilities from ALEPH, DELPHI and OPAL are discussed in Sect. 7.1.3. Vertex charge is the simple P_{PS}-weighted sum over all tracks associated with the vertex i.e.

$$Q_{\text{vtx}} = \sum_i q_i \cdot P_{\text{PS},i} \tag{5.2}$$

where q_i is the measured charge of track i. The vertex charge is therefore a decay charge estimator but this variable was also developed to more optimally exploit the correlations from the tracks originating at the primary vertex. The construction is closely related to the jet charge,

$$Q_{\text{Pvtx}} = \frac{\sum_i q_i\,|p_i \cdot T|^{\kappa}\,(1 - P_{\text{PS}})}{\sum_i |p_i \cdot T|^{\kappa}\,(1 - P_{\text{PS}})}. \tag{5.3}$$

This form of tag is useful for neutral states where the application of decay flavour tags would on average return zero and hence provide no quark flavour discrimination. This tag was specifically developed for the production flavour tagging of B^0_d states.

5.1.3 Jet Charge and $BR(b \to \ell^- \bar{\nu}_\ell X)$

ALEPH [2], DELPHI [3] and OPAL [4] all exploited to good effect flavour tags in their measurements of BR^b_{sl}. Similar techniques were applied to the lepton samples used for the measurement A^b_{FB} which is discussed further below in Sect. 5.3. In Sect. 4.1 we saw that a lack of discrimination power between the lepton sources $(b \to \ell)$, $(b \to c \to \ell)$ and 'background' was a major stumbling block to extracting reliable BR^b_{sl} measurements in the early days of LEP. Production flavour tags such as jet charge can help since leptons from $b \to \ell$ decays have the same charge sign as the weakly decaying b-quark whereas leptons from $b \to c \to \ell$ decays have the opposite sign (see Fig. 5.2). Furthermore, the opposite hemisphere carries charge information that can be used since the charge correlation of a production flavour tag from the opposite hemisphere and the lepton, is opposite to that between the same-side flavour tag and the lepton.[1]

The OPAL analysis combined, via a neural network, the usual lepton kinematic variables of p and p_t (discussed in Sect. 4.1) with two extra classes of variable: (1) jet charge of the jet containing the lepton and of the most energetic jet in the event hemisphere opposite the lepton ; (2) lepton jet energy and lepton impact parameter significance which helped further in distinguishing $(b \to \ell)$ from $(b \to c \to \ell)$. In fact further optimisation was possible by firstly forming separate networks to recognise $b \to \ell$ and $b \to c \to \ell$ from all other types of event and secondly by training each network separately for electron and muon samples – motivated by the fact that the background is different in each. The distribution of both neural network outputs for muons are shown in Fig. 5.3. The semi-leptonic branching ratio BR^b_{sl} is extracted by fitting simultaneously the two network distributions for the fraction

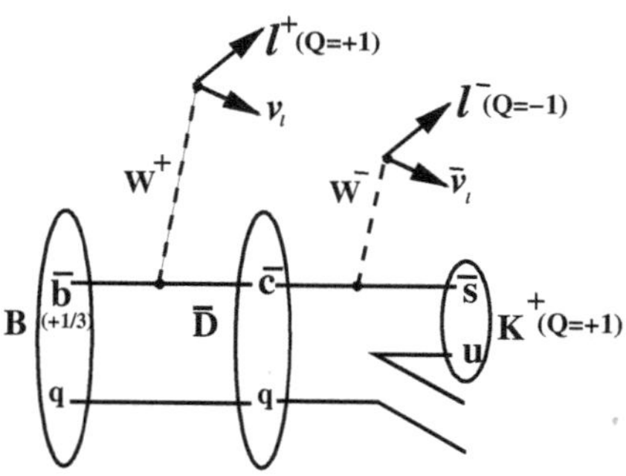

Fig. 5.2 The various charge correlations present in semi-leptonic B-meson decay chain between the lepton/kaon decay products and the parent $\bar{b}$ quark

[1] Note that with B^0_d-mesons these correlations become diluted due to $B^0 - \bar{B}^0$ mixing which transforms a b-quark to a $\bar{b}$ quark, and vice-versa, at some time between production and decay.

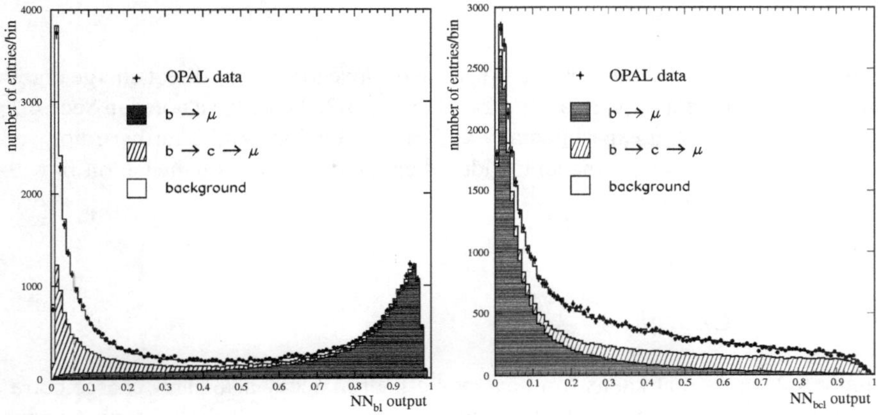

Fig. 5.3 (*left*) The output of a neural network trained to distinguish $b \to \ell$ events from all other sources of lepton (*right*). The output of a neural network trained to distinguish $b \to c \to \ell$ events from all other sources of lepton. From [4]

of $b \to \ell$ and taking the shapes of the direct, cascade and background components from simulation.

As soon as this new breed of LEP measurement was possible the discrepancy with the results from the $\Upsilon(4s)$ began to diminish as was shown in Fig. 4.1. The question of whether the semi-leptonic branching ratio as measured at LEP and the $\Upsilon(4s)$ are compatible, (a) with each other and (b) with the theoretical expectation, is a long-standing area of debate. The anxiety is justified since the semi-leptonic process $b \to c\ell\nu$ is free from much of the uncertainties associated with hadronic processes and failure to model it accurately would signal a failure of the whole heavy quark expansion approach. The consensus now is that there is consistency between measurements and the question of theoretical understanding is usually discussed in the context of charm counting. We will re-visit this topic in Sect. 7.3.

5.2 Decay Flavour

Decay flavour refers to the b-quark flavour present in the b-hadron at the time of the weak decay of this state. Due to neutral B-meson mixing, this flavour need not be the same as the b-quark flavour at production time i.e. directly after fragmentation. In contrast to the tracks correlated to production flavour, these particles originate from displaced secondary vertices but would also be expected to travel in roughly the same direction as the underlying b-quark. We now run through the most common reconstructed quantities that were exploited for decay flavour tagging at LEP:

5.2.1 Lepton Charge

Leptons produced in semi-leptonic b-hadron decay carry a very direct charge correlation with the parent b-quark as illustrated in Fig. 5.2. As we have noted in Sect. 4.1, leptons are very clean experimentally and the correlation can be further improved by tighter requirements on the lepton identification and/or cuts on the lepton p_t with respect to the jet axis.

5.2.2 Kaon Charge

Figure 5.2 shows that kaons from cascade c-hadron decay also carry charge correlation information. The power of this method is naturally highly dependent on the quality of kaon ID and on unravelling this source of kaons from other possibilities such as the associated production with a B_s^0 meson shown in Fig. 5.1.

5.2.3 Weighted Secondary Vertex Charge

A weighted secondary vertex charge variable is defined by replacing $(1 - P_{PS})$ in (5.3) with P_{PS}. The value of κ must be separately tuned in order to give the optimal charge separation. There is clearly an overlap between this variable and conventional vertex charge, but the idea here is to target more the correlation information in the leading particles from b-hadron decay by including a momentum component to the overall charge weight.

5.2.4 Vertex Charge

As defined in (5.2) a vertex charge summed over all tracks in an event hemisphere will return a value close to $+1$ (-1) in the presence of a charged b-hadron so tagging the hemisphere as containing a $\bar{b}$ (b) quark, whilst a vertex charge close to zero indicates a neutral b-hadron and gives no information on the b-quark production flavour. Based on binomial statistics, an uncertainty on Q_{vtx} can be defined as

$$\sigma_{Q_{\text{vtx}}} = \sqrt{\sum_i P_{\text{PS},i}\left(1 - P_{\text{PS},i}\right)}. \tag{5.4}$$

Clearly a vertex charge with large $\sigma_{Q_{\text{vtx}}}$, cannot distinguish between charged and neutral b-hadrons, and provides no information on the b-quark production flavour.

5.2.4.1 Extracting b-Hadron Production Fractions

A nice application of the vertex charge as described above was a first measurement of the production fractions of neutral and charged b-hadrons (see Sect. 1.3.7 for definitions) by the DELPHI collaboration [5, 6, 7]. Measurements using exclusive decays are very difficult since there are many decay channels with poorly known, small, branching ratios. Using vertex charge DELPHI were able to achieve a beautiful separation of charged from neutral states (see Fig. 5.4) which was fitted to the functional form,

$$F(Q) = \left(f_{X_B^+} \cdot Q_{X_B^+} + f_{X_B^0} \cdot Q_{X_B^0} \right) \tag{5.5}$$

Here, X_B^+ and X_B^0 indicates any charged and neutral b-hadron and by imposing the unitarity constraint, $\left(f_{X_B^+} + f_{X_B^0} \right) = 1$, the result of the fit was

$$f_{X_B^+} = (f_{\mathrm{B_u}} + f^+_{b-\mathrm{baryon}}) = 42.09 \pm 0.82(\mathrm{stat.}) \pm 0.89(\mathrm{syst.})\%$$

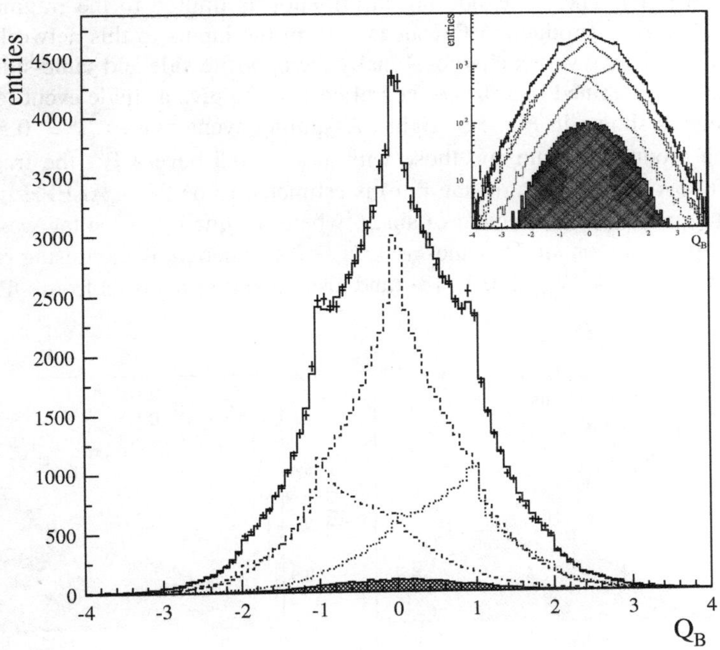

Fig. 5.4 The vertex charge distribution for data (*points*) with the result of the fit superimposed (*solid histogram*) on both a linear and log scale (*insert*). The distributions for neutral (*dashed histogram*), negatively (*dashed-dotted histogram*) and positively (*dotted histogram*) charged b-hadrons are also shown

5.3 Combined Flavour Tagging

The flavour tag tools described in Sect. 5.1 were combined in various combinations and used in various guises by the LEP collaborations according to the needs of the particular analysis task. The state-of-the-art is well illustrated by the ALEPH study of the CP asymmetry in $B^0 \to J/\Psi K^0_s$ where it is essential to know whether the reconstructed signal originated from a $\bar{B}^0$ or B^0_d at production. To do this the (same-side) hemisphere containing the signal B^0_d and the opposite hemisphere are considered separately and flavour tags are formed excluding the lepton and pion pairs that are the final decay products of the signal B^0_d. The production flavour can then be tagged on the same-side via the charge correlation with the fragmentation tracks and indirectly using the opposite side to tag flavour of the b-hadron produced in conjunction with the signal B^0_d.

For the opposite-side tag, essentially all of the flavour tools are input to a neural network: jet charge variables, vertex charge, weighted primary and secondary vertex charge, lepton charge and kaon charge. In addition, use is made of quality variables including the vertex charge uncertainity from (5.4) and the decay length of the signal B^0 which 'informs' the network about how large the separation is between the B^0 decay and fragmentation systems. The output of this network, x_{opp}, is shown in Fig. 5.5 (left). The same-side tag information is limited to the fragmentation tracks and so only production flavour tags form the inputs to this network i.e. jet charges and primary vertex charges. Finally the opposite-side and same-side neural networks are combined in a further neural network to give a single event-level tag, x_{tag}, which is shown in Fig. 5.5 (right). Assuming events with $x_{\mathrm{tag}} > 0.5$ have a B^0_d in the production state and those with $x_{\mathrm{tag}} < 0.5$ have a $\bar{B}^0$, the fraction of incorrect tags (the average mistag rate) is estimated to be 28%. ALEPH used this flavour tag in their measurement of $\sin 2\beta$ where the quality of the tag was crucial since the precision on $\sin 2\beta$ scales as $1/(1 - 2\omega)$ where ω is the mistag rate. The result $\sin 2\beta = 0.84^{+0.82}_{-1.04} \pm 0.16$ [8] (and also an earlier analysis from OPAL [9])

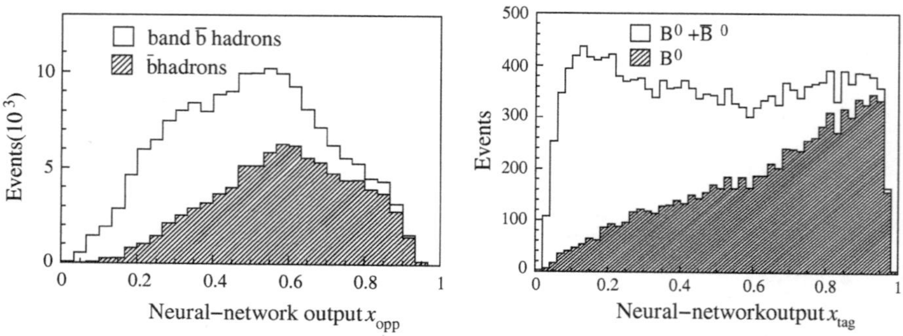

Fig. 5.5 Neural Network output distributions for simulated signal events: (*left*) opposite-side production flavour tag with the contribution from hemispheres containing $\bar{b}$-hadrons shaded; (*right*) event level production flavour tag with the contribution from B^0_d decays shaded

was however completely dominated by statistical uncertainties related to the small sample of exclusive decays $B^0 \to J/\Psi K_s^0$. The issue of exclusive reconstruction is discussed further below in Sect. 7.2.

An important analysis where progress was intimately linked with developments in flavour tagging was the measurement of the forward-backward asymmetry for b-production introduced in Sect. 1.3.2. The final publications from the LEP collaborations were based on combined flavour tags, such as described above for the case of the ALEPH tag, but with the lepton charge variables removed. This was important to reduce the level of correlation between the A_{FB}^{b} inclusive analyses (ALEPH [10], DELPHI [11], L3 [12] and OPAL [13]) and those based on leptonic samples (ALEPH [14], DELPHI [15], L3 [16] and OPAL [17]) enabling a better final precision to be attained on A_{FB}^{b} by combining the inclusive and leptonic measurements. This topic is addressed further in Sect. 6.4.

References

1. R.D. Field, R.P. Feynman, Nucl. Phys. B **136**, 1 (1978)
2. ALEPH Collab., A. Heister et al.: Eur. Phys. J.C **22**, 613 (2002)
3. DELPHI Collab., P. Abreu et al.: Eur. Phys. J.C **20**, 455 (2001)
4. OPAL Collab., G. Abbiendi et al.: Eur. Phys. J.C **13**, 225 (2000)
5. DELPHI Collab., J. Abdallah et al.: Phys. Lett. B **576**, 29 (2003)
   rates and decay properties. CERN-EP/2001-050
6. ALEPH Collab., D. Buskulic et al.: Phys. Lett. B **361**, 221 (1995)
7. ALEPH Collab., R. Barate et al.: Eur. Phys. Lett. J. C **2**, 197 (1998)
8. ALEPH Collab., R. Barate et al.: Phys. Lett. B **492**, 259 (2000)
9. OPAL Collab., K. Ackerstaff et al.: Eur. Phys. J. C **5**, 379 (1998)
10. ALEPH Collab., A. Heister et al.: Eur. Phys. J. C **22**, 201 (2001)
11. DELPHI Collab.: Determination of A_{FB}^{b} at the Z pole using inclusive charge reconstruction and lifetime tagging., CERN-EP/Paper 329 (2003) (submitted to EPJ)
12. L3 Collab., M. Acciarri et al.: Phys. Lett. B **439**, 225 (1998)
13. OPAL Collab., G. Abbiendi et al.: Phys. Lett. B **546**, 29 (2002)
14. ALEPH Collab., A. Heister et al.: Eur. Phys. J. C **24**, 177 (2002)
15. DELPHI Collab., J. Abdallah et al.: Eur. Phys. J. C **34**, 109 (2004)
16. L3 Collab.,M. Acciarri et al.: Phys. Lett. B **448**, 152 (1999)
17. OPAL Collab., G. Abbiendi et al.: Phys. Lett. B **577**, 18 (2003)

Chapter 6
Double-Hemisphere Tagging

6.1 Introduction

In the bid to make precision measurements that depend on simulations, an analysis will eventually be limited by the uncertainties associated with the simulation not being a perfect model of the data. Double hemisphere techniques were employed quite widely across LEP b-physics analyses as a way of extracting tagging efficiencies and sample purities direct from the data without depending on simulation. The basic topology of $Z^0 \to b\bar{b}$ events (illustrated in Fig. 1.3), where the quark and anti-quark can be considered as separating back-to-back into opposite hemispheres and evolving as essentially independent systems, lends itself naturally to this type of analysis.

Measuring the $Z^0 \to b\bar{b}$ event fraction R_b (defined in Sect. 1.3.3) was one of the primary drivers behind the development of the double hemisphere technique. The analysis is basically a simple counting experiment based on a b-quark tag which could be any of the methods described in Chap. 4. The fraction of event hemispheres passing a cut in this tag is given by

$$F_S = \epsilon_b R_b + \epsilon_c R_c + \epsilon_{uds}(1 - R_b - R_c). \tag{6.1}$$

Here, ϵ_f is the efficiency to tag a hemisphere in events where the Z^0 has decayed to quark pairs of flavour f and the subscript 'S' refers to the fact we are tagging single hemispheres. In principle, 6.1 can be solved for R_b if all the efficiencies are estimated from simulation and R_c is set to the Standard Model prediction. If the tag is very good, the light quark efficiencies will be small and very little uncertainty will enter the measurement by estimating them from simulation. The most precise measurements are eventually limited by the uncertainties in estimating ϵ_b from a simulation which cannot model b-physics and detector effects perfectly. If however we now measure the fraction of events in which *both* hemispheres are tagged and if the hemispheres are independent of each other, the fraction of double-tagged events is

$$F_D = \epsilon_b^2 R_b + \epsilon_c^2 R_c + \epsilon_{uds}^2(1 - R_b - R_c). \tag{6.2}$$

G.J. Barker, *b-Quark Physics with the LEP Collider*, STMP 236, 121–128,
DOI 10.1007/978-3-642-05279-8_6, © Springer-Verlag Berlin Heidelberg 2010

We now have two equations with effectively two unknowns, R_b and ϵ_b, which can be solved for – and all other parameters are estimated from the simulation.

There are two caveats to the use of such double-tagging methods. The first is that the statistical precision on the measured parameters is limited by the statistics of the double-tagged events. The number of events needed to get the same statistical precision as the single-tag method is proportional to $1/\epsilon_b^2$ and so use is restricted to tags that are highly efficient. The second caveat concerns the assumption that the hemispheres are independent which is only approximately true. In reality the hemisphere tags are always correlated due to one or a combination of the following factors:

1 b-hadrons are produced in jets roughly back-to-back and detectors tend to be symmetrical around the interaction point. Therefore if the tag is reconstructed in a region of poor geometrical acceptance on one side the same will be true on the other side and both hemispheres will be relatively poorly tagged.
2 Hard gluon emission will tend to reduce the momentum of both primary quarks so impacting on the tag of both hemispheres in the same way.
3 If a common primary vertex is used for the decay length calculation of both hemispheres (either averaged over many events or fitted per event), its resolution will impact on the likelihood of lifetime tagging both hemispheres in the same way.
4 A b-hadron with a particular lifetime will have a long decay length if it hadronises with a large momentum and the parent hemisphere will be efficiently tagged. In this case, there will be less primary tracks available for the primary vertex fit, so degrading its resolution and making it less likely that the opposite hemisphere will be tagged.

The presence of hemisphere correlations modifies the simple relationship between the single and double-tag efficiency as,

$$\epsilon_f^{\text{double}} = (1+\rho)\epsilon_f^2 \tag{6.3}$$

where $\rho \neq 0$ quantifies the level of hemisphere correlation and can be positive or negative. The last two sources of correlation above can be eliminated by using a primary vertex point for each hemisphere reconstructed from only the particles associated with that hemisphere (see Sect. 4.4.2 for primary vertex reconstruction). Since the efficiency for reconstructing two primaries per event can be low due to lack of tracks, the trade off between statistical precision and having lower correlations must be carefully considered.

6.2 Double-Tag Analyses

Use of high performance b-tags and hemisphere correlations that were in general rather small (and so could be evaluated with the simulation with low impact), meant that double-tag techniques were an important factor in controlling systematics on many of the most precise b-physics results from LEP. In the drive to push the precision on R_b ever smaller the technique was extended with great success to a multi-tag method which is discussed further in Sect. 6.3 below. Almost all of the final high-statistics b-physics analyses made use of the gain in systematic understanding that the double hemisphere tag gave. Other examples include measurements of the mean b-hadron lifetime [1] and the semi-leptonic branching fractions that extracted the purity of the $Z^0 \to b\bar{b}$ sample from the data [2].

Along with R_b, the other major beneficiaries of double-tag methods were the b-quark forward-backward asymmetry analyses. Precision measurements of A_{FB}^{b} ultimately placed the best constraint on $\sin^2\theta_{eff}^{\ell}$ from LEP (see Sect. 1.3.2). In general, a forward-backward asymmetry analysis of $Z^0 \to q\bar{q}$ events has two main requirements: (1) the event sample is enhanced in the quark flavour of interest with very low background and the fraction of this flavour (F_q) out of the total event sample must be known; (2) the efficiency of tagging correctly the quark charge (ϵ_q) is known and is as high as possible. The asymmetry that is measured is then related to the physics quantity by [3],

$$A_{FB}^{meas} = \sum_q (2\epsilon_q - 1) F_q A_{FB}^{q\bar{q}}. \tag{6.4}$$

Note that if $\epsilon_q = 50\%$, the measured asymmetry is zero which is sensible since tagging the quark charge correctly only 50% of the time is equivalent to random guessing and there is no sensitivity to an underlying asymmetry. Of-course the flavour fraction and the charge tag efficiency could be estimated from simulation but double-tag techniques allow both to be extracted direct from the data. The fractions follow by comparing single to double-tag rates in a way similar to that discussed above for R_b but there is also information about the charge tag efficiency. The charge tagging can only be correct if the sign of the tag flips between opposite event hemispheres since the quark and anti-quark carry opposite charges. In fact, for a sample of pure $Z^0 \to b\bar{b}$ events and ignoring hemisphere correlations, the fraction of same-sign double tags out of the sample of all double tags is given by [3],

$$F_{SS} = 2\epsilon_b(1 - \epsilon_b). \tag{6.5}$$

This form is believable since if we always tag the charge correctly, $\epsilon_b = 100\%$ and the double charge tags will always be opposite sign (i.e. $F_{SS} = 0$) and if we tag charge correctly only 50% of the time ($\epsilon_b = 50\%$) the double tags will only be correct 50% of the time (i.e. $F_{SS} = 50\%$). The double-tag relationship (6.5) can therefore be used to solve for the unknown charge tag efficiency up to (small) corrections for background and hemisphere correlations for which the simulation can be used.

By evaluating tagging efficiencies and sample purities directly from the data, double-tag methods were also commonly implemented as part of a calibration correction for simulation to better match the data. DELPHI's combined charge tag A_{FB}^{b} analysis [4] used the double tag methods described above to extract both the

$Z^0 \to b\bar{b}$ and $Z^0 \to c\bar{c}$ event flavour tagging efficiencies as a function of various cuts on the flavour tag. These efficiencies were then evaluated from the simulation and data and then made to agree by shifting (or calibrating) the flavour tag variable in the simulation. The effect of cutting on the flavour tag over a wide range then agreed between data and simulation at the $\pm 1\%$ level and it was found that the combined effects of a whole swathe of uncertain b-physics effects, such as B^0_d mixing and B branching ratios could be accounted for in a single calibration correction.

Charge tags have also been calibrated against data by exploiting the fact that the two hemispheres in a $Z^0 \to b\bar{b}$ event should, up to small corrections, contain either a b or $\bar{b}$ quark which carry opposite charges. The number of opposite sign double tags is therefore related to the efficiency to tag the quark charge correctly which can therefore be calibrated by correcting the charge tag until the number of such double tags agrees between data and simulation. Examples of analyses that apply such a procedure can be found in DELPHI's measurement of the b-quark branching fractions into charged and neutral b-hadrons [5] and also in the analyses of A^b_{FB} and A^c_{FB} using prompt leptons [6]. An illustration of the high level of control possible after such calibration procedures is given in Fig. 6.1 from the lepton analysis which illustrates the wonderful level of agreement between data and simulation achieved over a wide range of the charge tag.

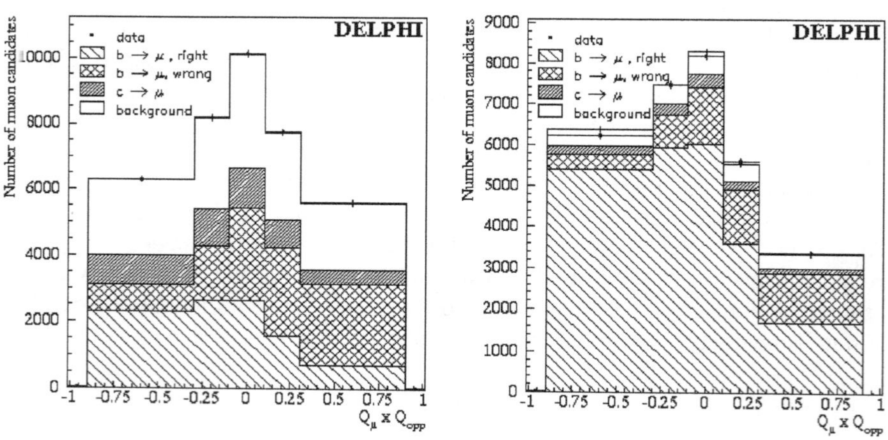

Fig. 6.1 Distributions of the jet charge product from both hemispheres of a sample of $b \to \mu$ events. The muon candidates have (*left*) $p_T < 1.3\,\text{GeV}/c$ and (*right*) $p < 7\,\text{GeV}/c$. From [6]

6.3 Multi-Tag Methods

The double tag formalism defined in (6.1) and (6.2) based on a single b-tag, can be extended to be based on multiple mutually exclusive tags so that each hemisphere is counted to be tagged by one and only one of the tags. These extra tags could be additional complimentary b-tags and/or tags that specialise in identifying the background to a particular analysis e.g. charm tags and u,d,s-tags. These ideas were investigated first in [7] and found application in the most precise R_b measurements to come out the LEP era: ALEPH [8], DELPHI [9]. The advantages of such an approach for the R_b analysis are:

(i) *All* hadronic hemispheres are tagged either as signal, background or unusable which helps to boost the statistical precision.
(ii) This in turn enables the b-tag to be operated with tighter cuts (i.e. at higher purity) than before so boosting sensitivity to R_b.

The fraction of double tags is now given by

$$F_{ij} = \sum_q (1 + \rho_{ij})\epsilon_i^q \epsilon_j^q R_q \tag{6.6}$$

which represents the fraction of events tagged by tag i in one hemisphere and by tag j in the other hemisphere where the sum goes over all quark flavours q. With N_T separate hemisphere tags the number of double tag fractions is, $N_T(N_T + 1)/2$. Any hemisphere left untagged is separately counted in a null-tag category so that all hadronic hemispheres are exclusively classified and leads to the following conditions being satisfied,

$$\sum_i \epsilon_i^q = 1 \tag{6.7}$$

$$\sum_i \epsilon_i^q \epsilon_j^q \rho_{ij}^q = 0 \tag{6.8}$$

where $q = uds, c, b$ and $i, j = 1, ..., N_T$. The first of these conditions tells us that not all tags are independent and for 3 different event types (uds, c, b) the number of unknown efficiencies is $3(N_T - 1)$. Since $\sum_q R_q = 1$ there are only 2 independent partial-width ratios and if all the hemisphere correlations derive from simulation the number of unknowns is $3(N_T - 1) + 2$ from $N_T(N_T + 1)/2 - 1$ independent double tag rates. From this, the minimum number of tags needed to over constrain the system is six and this is the number of tags that the ALEPH and DELPHI analyses were based on. However, it was known from the studies in [7] that ambiguities in this system of equations make it impossible to solve for the full array of unknowns. The problem is only solvable by constraining more of the unknowns and their correlations, to their predictions from simulation. In the LEP analyses this meant fixing R_c (and explicitly accounting for variations in this value in the systematics) and taking the background efficiencies $(\epsilon_c, \epsilon_{uds})$ from simulation for the best b-tag available – since in this case the background efficiencies were the smallest entering the analysis. Although all correlations are calculated from simulation, the second condition (6.8) allows one to be constrained by the others and the natural choice for this is the null-tag correlation which has a potentially complex flavour mix and hence potentially large

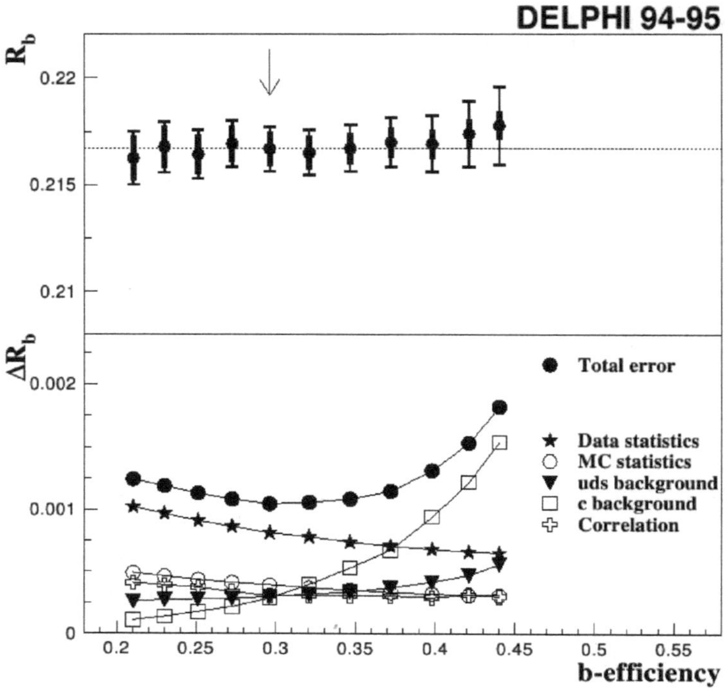

Fig. 6.2 R_{b} results versus the (best) b-tag efficiency from the DELPHI multi-tag analysis. The thick (thin) error bars are statistical (total) and the *arrow marks* the minimum in the total error at $\epsilon_b = 29.6\%$ at which point the purity in $Z^0 \to b\bar{b}$ events was 98.5%. From [9]

systematic uncertainty. In general the introduction of extra (perhaps complicated) tags does not inflate the systematic uncertainty if careful choices are made regarding which quantities to evaluate from the data and which to take from the simulation.

Figure 6.2 shows the R_{b} measurements from the DELPHI analysis as a function of the efficiency for the best performing b-tag used. This measurement was the single most precise from the LEP era and the stability over a large range of efficiencies is remarkable.

6.4 The $A_{\mathrm{FB}}^{\mathrm{b}}$ 'Problem'

The most precise measurements of $A_{\mathrm{FB}}^{\mathrm{b}}$ from the LEP collaborations [10, 4, 11] ultimately came from analyses using rather complex combined charge tags that were carefully selected to be uncorrelated with lepton-based charge tags. These results could then be optimally combined with uncorrelated measurements of $A_{\mathrm{FB}}^{\mathrm{b}}$ based on lepton tags by the EWWG prescription (see Sect. 1.4). There were slight statistical gains to be had by running the analyses in bins of $\cos\theta$ and fitting for $A_{\mathrm{FB}}^{\mathrm{b}}$ based on the following angular dependence predicted by the Standard Model (see e.g. [10]),

$$A_{FB}^{b}(\cos\theta) = \frac{8}{3} A_{FB}^{b} \frac{\cos\theta}{1+\cos^2\theta}. \tag{6.9}$$

The application of double-tag techniques helped reduce the systematic uncertainties leaving the results statistically limited and with final relative uncertainties at the few percent level. With this level of precision inconsistencies with the asymmetry data, which had been a feature of the early LEP results but at low significance, could not be ignored.

The size of the problem for measurements of A_{FB}^{b} and $\sin^2\theta_{eff}^{\ell}$ is illustrated in Fig. 6.3. Combined data from LEP on A_{FB}^{b} have been consistently below the Standard Model expectation ever since the first publications and based on the final LEP results, the value of $\mathcal{A}_b$ (calculated by using 1.7) was 3.2 standard deviations below the Standard Model expectation [3]. Similarly, values of $\sin^2\theta_{eff}^{\ell}$ extracted from measurements of A_{FB}^{b}(LEP) and A_{LRFB}^{b}(SLD) based on the Standard Model relationships 1.7, 1.8 and 1.10 are also currently in disagreement at about the three-standard deviation level. In fact both of these discrepancies are manifestations of the same problem and can be traced back to the effective coupling of the Z^0 to the b-quark [3, 12]. Global fits to all of the heavy flavour electroweak data to extract the most likely values of the couplings, (g_A^b, g_V^b) defined in (1.2) and (1.3), are found to deviate from the Standard Model expectation by about two standard deviations. Further, if analysed in terms of the left- and right-handed effective couplings, $g_L^b = (g_V^b + g_A^b)/2$ and $g_R^b = (g_V^b - g_A^b)/2$, it is found that essentially all of the discrepancy is contained in g_R^b alone.

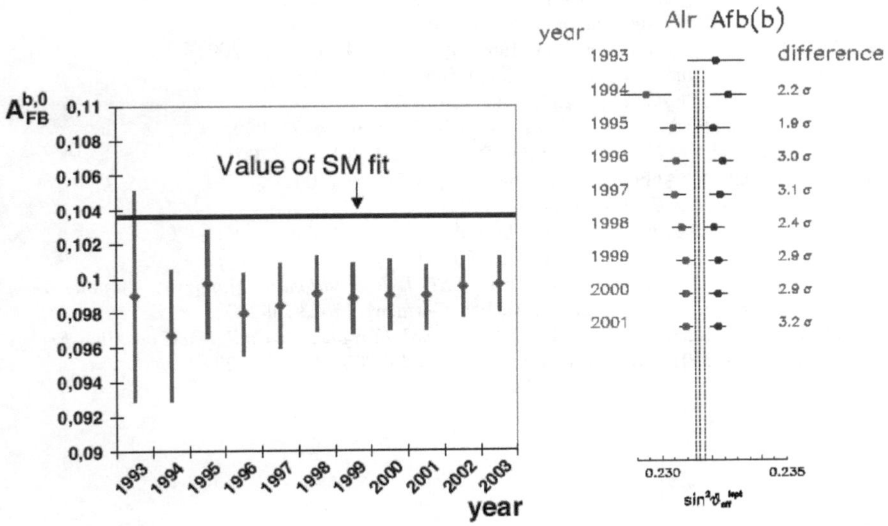

Fig. 6.3 The history of the discrepancy between measurements of (*left*) $A_{FB}^{0,b}$ (from [15]) and (*right*) $\sin^2\theta_{eff}^{\ell}$ extracted from SLAC's left–right polarisation asymmetry and LEP's forward backward b-asymmetry (from [16]). Note that the pole asymmetry A_{FB}^{b} is labelled $A_{FB}^{0,b}$ in this plot

A non-Standard Model value of the right-handed coupling to b-quarks can therefore accommodate the b-quark asymmetry results from LEP but why one coupling should fall outside the Standard Model picture remains a mystery. Attempts have been made to explain these results in terms of statistical fluctuations or failings with systematic uncertainty studies (e.g. [13]). Inevitably there have also been suggestions that new physics is responsible (e.g. [14]). The resolution of the problem will likely have to wait for the next generation of e^+e^- collider and passes into the LEP b-physics legacy.

Finally, it is interesting to note that $A^{\mathrm{b}}_{\mathrm{FB}}$ is responsible for the largest χ^2-contribution in the fit to the electroweak data set that predicts the mass of the Higgs boson. Furthermore, if the b-asymmetry is left out of the fit, the resulting Higgs mass falls below (albeit with large uncertainty) the current lower limit of 114 GeV set by the LEP direct search! This is further evidence that the b-asymmetry results look inconsistent with the rest of the electroweak data set and suggests using extreme caution in adopting values of the Higgs mass prior to the resumption of the direct search at the LHC.

References

1. OPAL Collab., K. Ackerstaff et al.: Zeit. Phys. C **73**, 397–408 (1997)
2. OPAL Collab., G. Abbiendi et al.: Eur. Phys. J. C **13**, 225 (2000)
3. ALEPH, DELPHI, L3, OPAL, SLD, LEP Electroweak Working Group, SLD Electroweak Group and SLD Heavy Flavour Group: Phys. Rep. **427**, 257 (2006)
4. DELPHI Collab., J. Abdallah et al.: Eur. Phys. J. C **40**, 1–25 (2005)
5. DELPHI Collab., J. Abdallah et al.: Phys. Lett. B **576**, 29 (2003)
6. DELPHI Collab., J. Abdallah et al.: Eur. Phys. J. C **434**, 109–125 (2004)
7. P. Billoir et al.: Nucl. Instrum. Methods A **360**, 532–558 (1995)
8. R. Barate et al.: Phys. Lett. B **401**, 163–175 (1997)
9. DELPHI Collab., P. Abreu et al.: Eur. Phys. J. C **10**, 415–442 (1999)
10. ALEPH Collab., A. Heister et al.: Eur. Phys. J. C **22**, 201–215 (2001)
11. OPAL Collab., G. Abbiendi et al.: Phys. Lett. B **546**, 29 (2002)
12. P. Renton: Eur. Phys. J. C **8**, 585–591 (1999)
13. J.H. Field, D. Sciarrino: Mod. Phys. Lett. A **15**, 761–774 (2000)
14. M. S. Chanowitz: Phys. Rev. Lett. **8**, 231802 (2001)
15. M. Elsing: *Heavy Flavour Results from LEP 1. International Europhysics Conference on High-Energy Physics* (HEP 2003), Aachen, Germany, 17–23 Jul 2003.
16. W. Venus: A LEP Summary. In: *International Europhysics Conference on High-Energy Physics* (HEP 2001), Budapest, Hungary, 12-18 Jul 2001, hep2001/284

Chapter 7
Optimal b-Flavour and b-Hadron Reconstruction

Without the luxury of a fully reconstructed b-hadron state, to make progress in b-physics, one is at the mercy of detector design and performance. As we have seen, precision tracking and a particle identification capability are prerequisites to be able to reconstruct the underlying physics event from the particles that make it through the detection-reconstruction chain. The experimenter must use the detector information available to make a selection of the particles that are most likely to be involved in the process of interest. This selection procedure could take the form of simply cutting away those particles whose parameter values fall outside of predetermined bounds. This cut-based approach has been traditionally the main technique of data analysis in high energy physics and its simplicity can produce measurements that are well 'understood' i.e. with a low systematic uncertainity. However, the performance of this method is best when applied to variables that (a) provide excellent discrimination and (b) are uncorrelated. This follows since cutting on a variable that shows only a weak discrimination will lead to a low selection efficiency and high background, whereas highly correlated variables bring no extra information to an analysis and so cannot improve the performance. Furthermore, rejecting particles from an analysis in this way assumes that they are all independent of each other which is often not the case e.g. measurements of track impact parameters, referenced to the reconstructed primary vertex position, are correlated between those tracks that formed part of the primary vertex fit. For these reasons, cut-based analyses tend to be restricted to the use of a few, high performance, variables that have small correlations between them.

The inclusive methods developed at LEP were far removed from the cut-based approach and were driven by the need to have tags with the best possible performance but retaining a high efficiency. The main conclusion of the LEP studies was that more optimal tags could often be formed by basing them on *all* reconstructed particles in the event and utilising a far larger scope of discriminating variables than the typical cut-based approaches. These developments are the subject matter of this chapter.

G.J. Barker, *b-Quark Physics with the LEP Collider*, STMP 236, 129–162,
DOI 10.1007/978-3-642-05279-8_7, © Springer-Verlag Berlin Heidelberg 2010

7.1 Inclusive b-Physics Tools

The basic idea behind inclusive tags is that the discriminating variables are combined to form a probability for each particle to be correlated to the physics process or quantity of interest and this probability can then be used as a weight to form the tag at the event or hemisphere level. As seen in Chap. 5, an example of this is the reconstruction of the b-hadron charge by first constructing a probability (P_i) for each particle i to be a decay product of the state and then defining the charge as the weighted sum $Q = \sum_{i=1}^{n} Q_i \cdot P_i$ over all n tracks of charge Q_i. We now discuss some of the analysis tools and reconstructed quantities that were important in constructing such inclusive probabilities.

7.1.1 *Artificial Neural Networks*

Since a major requirement of the new inclusive methods is to combine correlated information, neural networks were used extensively – especially in the latter years of LEP data analysis. A vast literature exists on the subject of neural networks which we will not attempt to repeat here – for a good introduction to the use of networks in high energy physics see e.g. [1]. In fact experience has shown that, in almost all cases, good results from high energy physics data have been possible with neural networks in rather standard configurations: feed-forward networks with a three-layer architecture (input layer, hidden layer and output layer) where the hidden layer contains a few more nodes than the input layer. A sigmoid 'activation function'[1] is most commonly used to determine the output of each node and the favoured scheme of network 'training' is via the back-propagation method. In general, studies have shown that no great improvements in the results can be gained by moving to more complex network configurations and often the simpler networks are found to be easier to train and quicker to train. At least part of the reason for this conclusion is that the use of neural networks in particle-physics analysis is still relatively new and the list of tasks to which they have been applied is short. By far the most common application is the relatively simple use of a network to make a classification decision e.g. is this event a b-event or not? An example of a more complex application for a network is mentioned later in Sect. 7.2.4 with respect to reconstructing the b-hadron energy.

Experience of using networks in analysis has identified some areas where care is needed to obtain the best results:

- The selection of discriminating variables that have simultaneously a high correlation to the target value and a low correlation with each other.
- Since simulation must almost always be used to train the networks on inclusive b-physics quantities, choosing discriminating variables that are well modelled is

---

[1] This is a 'S'-shaped turn-on function, $\frac{1}{1+\exp^{-Ax}}$ where x is the sum of all input weights to the node and A is a tuned parameter controlling the sharpness of the turn-on.

as important for this technique as it is for the likelihood method. Further, because the way in which any individual input variable affects the network output is non-trivial, it is difficult to account for any mismatch between the simulation and real data as a systematic error on a final measurement. Sometimes situations arise where the total error on a measurement is improved by excluding a particular discriminating variable simply because of the systematic uncertainty introduced by its inclusion in the network.

- The pre-processing of input variables so that they are all presented to the network in the same numerical range e.g [0, 1] and the transformation of inputs into a space where they are independent of each other. These kinds of measures increase the chance that the network, in the training process, finds the global minimum in the comparison of the network output and the target value.
- The use of 'quality variables' that in themselves carry little or no discriminating power but which provide the network with information during training about the reliability of the input variables e.g. the decay length variable associated with a secondary vertex with a low fit χ^2-value is likely to be more reliable estimator than a decay length with respect to a poorly reconstructed vertex.

7.1.2 Rapidity

Rapidity is a variable frequently used to characterise the longitudinal content of particle motion in reactions. It is defined as,

$$y = \frac{1}{2} \ln \left(\frac{E + p_L}{E - p_L} \right) \tag{7.1}$$

for a particle with four-vector $(\boldsymbol{p}, E)$ where p_L is the longitudinal component of the momentum along a reference direction. Choosing the direction to be the axis of an associated jet or perhaps the vector joining the event primary vertex to a secondary vertex, the rapidity becomes a useful discriminant between tracks from b-hadron decay and those from other sources. Evidence of this was seen in Fig. 4.25(c) when considering the rapidity of particles that made up a secondary vertex with respect to the axis of the jet that these particles were associated with. The rapidity of particles from b-hadron decay was found to be softer than those from c-hadrons and the particles from u,d,s events were softer still. At first glance it may seem surprising that the rapidity content in b-hadron decays is softer than for c-hadrons since the average energy of b-hadrons is higher. However, the mass and hence the mean decay multiplicity in b-hadron decays is also higher resulting in a mean energy per particle that is lower than for c-hadrons which leads to a mean rapidity that is also lower. In u,d,s events, secondary vertices are induced by badly measured tracks of some kind since, apart from K^0_s, Λ decays, there is almost no lifetime content in the event and all particles originate at the primary vertex. Displaced tracks could be caused by large multiple scattering and/or interactions with the detector material and tend to be rather soft so that their rapidity distributions are shifted to very low values.

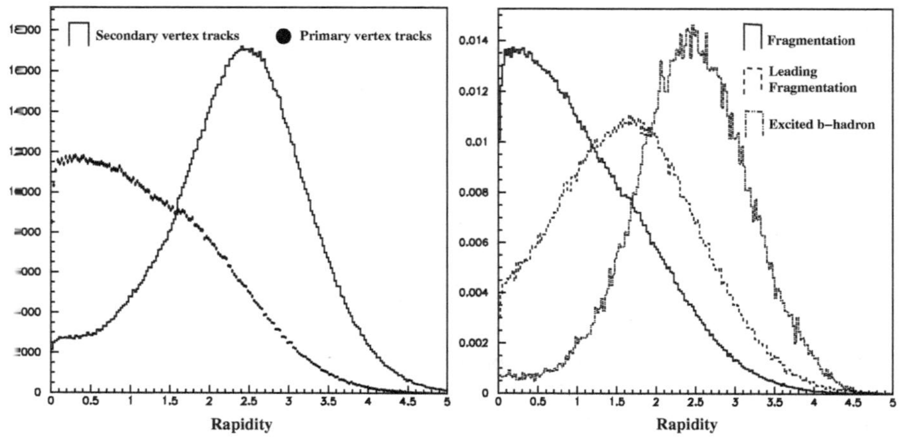

Fig. 7.1 (*left*) Comparing the rapidity distributions of tracks from the primary vertex, to tracks originating from some displaced secondary vertex. (*right*) Shows rapidity distributions of tracks originating from the fragmentation of a b-quark and from the decay of a b-hadron excited state

In addition to tagging the presence of b-hadrons, rapidity is also very useful in identifying the origin of particles in $Z^0 \rightarrow b\bar{b}$ events. Figure 7.1 shows rapidity distributions of tracks reconstructed from Monte Carlo $Z^0 \rightarrow b\bar{b}$ events that have been passed through the DELPHI detector simulation. The reference axis is the jet axis to which the tracks have been associated. The left-side plot illustrates the difference in rapidity distributions of tracks that originate from the primary vertex compared to tracks from any displaced secondary vertex such as the b-hadron decay point. A cut placed at about 1.5–2.0 in rapidity separates efficiently the two track types and could be used, for example, to reconstruct a b-hadron four-vector estimate from the tracks in a jet. This observation illustrates the fact that particles produced in the fragmentation process show much less of a correlation in direction with the b-hadron than do direct decay products of the b-hadron itself.

To take this one step further, the right-side plot of Fig. 7.1, shows how the rapidity can also be useful in identifying tracks of different origin from within the sample of tracks that come from the primary vertex. Tracks from the fragmentation process are significantly shifted to lower values compared to the decay products of b-hadron excited states which decay either strongly (B^{**}) or electromagnetically (B^*) before the state has a chance to move significantly from the primary vertex position. It is interesting to see that particles produced first in the fragmentation chain, the so-called leading fragmentation particles, are influenced enough by the direction of the b-quark momentum that their rapidity distribution is noticeably shifted to higher values compared to the remaining tracks produced in the fragmentation process.

Finally, it is worth noting that a Lorentz boost β along the b-hadron direction will add a constant $\ln(\gamma + \gamma\beta)$ to the particle rapidities and will leave rapidity differences unchanged. This leads to a nice property of rapidity – namely that the discrimination between b-hadron decay tracks and the background is maintained regardless of whether the particle momenta are calculated in the lab frame or in the rest frame of the b-hadron.

7.1.3 Primary-Secondary Track Probability, P_{PS}

An extremely useful quantity that found many applications in the final inclusive analyses from the LEP collaborations, was the concept of constructing a probability for *each track* to originate from a secondary vertex. To achieve this, various b-tag variables are combined, usually with a neural network, resulting in a probability (P_{PS}) that can be either cut on or used in the form of a weight. As seen in Chap. 4, one of the most successful applications of (P_{PS}) is to weight track charges to help in b-quark flavour tagging (i.e. b or $\bar{\text{b}}$) necessary for the measurement of e.g. A_{FB}^{b} and neutral B-meson oscillations. Here we describe briefly the different P_{PS} variables developed at LEP and some of the other applications.

The OPAL Collaboration developed a P_{PS} variable based on the following inputs to a neural network: momentum, $\cos\Theta$ (where Θ is the angle between a track and it's associated jet axis) and the impact parameter significance of the track ($S = d/\sigma_d$) with respect to the reconstructed primary and secondary vertex separately. The network was trained on tracks in b-jets to return the target value of 1 for tracks from the b-hadron decay chain and a value of 0 for fragmentation tracks. The network output is illustrated in Fig. 7.2 and shows a clear separation of the two track classes.

OPAL utilised the P_{PS} probability in a number of situations. By cutting on $P_{PS} < 0.5$ it was used to measure the fragmentation component of b-hadron energy in a measurement of τ_b [2]. By using P_{PS} instead of impact parameter probabilities, OPAL formed a mass tag similar to that developed by ALEPH (described in Sect. 4.5) and used it as part of an R_b measurement [3] and also in the $|V_{cb}|$-exclusive analysis [4] based on the decay mode $\bar{B}_d^0 \to D^{*+}\ell^-\bar{\nu}_\ell$.

The ALEPH P_{PS} variable, described in [5], was based on the following inputs to a neural network: the rapidity and longitudinal momentum relative to an estimate of the b-hadron flight direction, and the track probability $P(S)$, defined in (4.14), calculated with respect to a reconstructed secondary vertex. In addition, the network was presented with some quality variables related to how well the secondary vertex was reconstructed.

DELPHI has also developed a P_{PS} variable which played a central role in their inclusive b-physics package BSAURUS [6]. Input variables to the DELPHI neural network are:

- Track momentum.
- Track probability $P(S)$, defined in (4.14), calculated separately with respect to the reconstructed primary and secondary vertex.
- The momentum and decay angle of the track in the *centre-of-mass frame* of the b-hadron candidate. The b-hadron is reconstructed based on the rapidity algorithm outlined above by summing up all track 4-vectors in the jet with rapidities greater than a cut of 1.6. The expectation, illustrated in Fig. 7.3, is for the decay

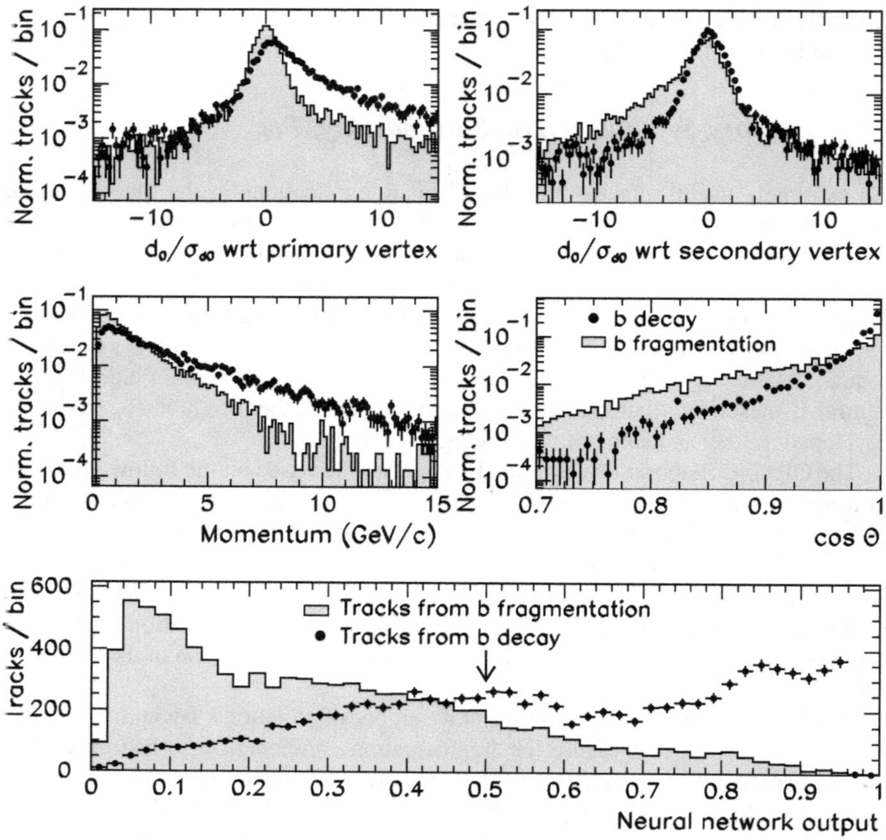

Fig. 7.2 The four input variables to the OPAL P_{PS} variable and the output distribution displaying separately tracks from b-hadron decay (*dots*) and tracks from b-fragmentation (*histogram*). From [2]

products of the b-hadron to be isotropically distributed in space (consistent with the decay of a spin zero particle) whereas the softer fragmentation tracks will show an angular distribution peaked in the direction back towards the primary vertex location.

- The track rapidity with respect to the jet axis that the track is associated with.
- A flag to signal whether the track was included in the secondary vertex fit.

DELPHI studies revealed that significant improvements in the network variable performance could be achieved by the inclusion of quality variables such as: (1) some measures of the possibility that a track had reconstruction errors e.g. the number of track 'hits' that are shared with other tracks in the event, (2) the decay length significance L/σ_L of the reconstructed secondary vertex, a quantity which is correlated to the vertex resolution and (3) secondary vertex mass, where values far from the nominal b-hadron mass flag that there is a good chance of the vertex

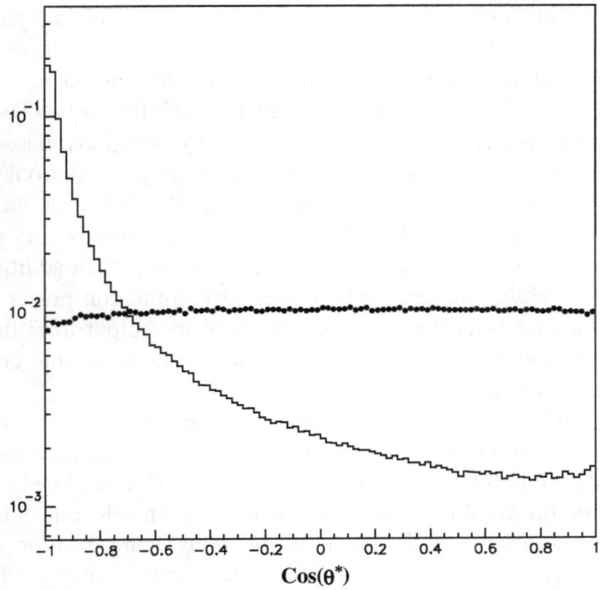

Fig. 7.3 In the b-hadron rest frame, the cosine of the angle between the track vector and the b-hadron direction is plotted for simulated $Z^0 \to b\bar{b}$ events. The points show tracks from direct b-hadron weak decay and the histogram shows tracks from the fragmentation process

being poorly reconstructed. The output of the DELPHI variable is shown in Fig. 7.4 (left) – note that this plot is logarithmic, in contrast to Fig. 7.2!. The relative amount of 'signal' (i.e. tracks from the b-hadron decay chain) and 'background' (i.e.other tracks in $e^+e^- \to Z^0 \to b\bar{b}$ events) in the distribution reflects the natural mix found

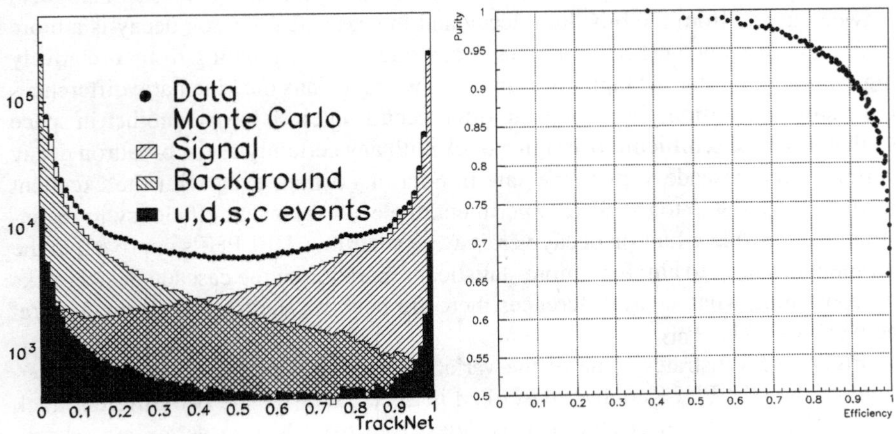

Fig. 7.4 (*left*) the DELPHI P_{PS} variable comparing data to the simulation. (*right*) The purity vs efficiency performance

in a sample of two-jet events, that have passed fiducial cuts and a b-tag selection to give a purity in $Z^0 \rightarrow b\bar{b}$ events of about 80%. If the signal and background had the same normalisation however, the histograms would cross at $P_{PS} \simeq 0.5$ representing the point at which tracks have an equal probability to be classed as either signal or as background. By making cuts successively harder and harder on the P_{PS} variable, the purity vs efficiency performance for tagging tracks from the b-hadron decay can be formed and is shown in Fig. 7.4 (right) – based on an independent sample to that used for training in order to avoid biases. Clearly very pure samples of b-hadron decay tracks can be isolated by this technique. In addition, the input variables were carefully chosen for both their discrimination power and realistic modelling of the data, with the result that the network output from the simulation is in good agreement with data and can be used with reasonable confidence for precision measurements.

DELPHI found that the efficiency for finding secondary vertices in b-hemispheres was improved by selecting tracks with large P_{PS} values in addition to the standard selection based on e.g. track displacement or particle ID as outlined in Sect. 4.4.3. It was found that the combined variable was efficient in selecting b-hadron decay tracks which none-the-less failed to pass the individual selection cuts. In fact because the separation power of some of the input variables improved with a more precise determination of secondary vertices, the final form of the DELPHI P_{PS} variable was a 'second iteration' of the original network variable. In the second iteration, all input variables that depended on the secondary vertex, were recalculated with the improved vertex track selection algorithm, and the network re-trained.

7.1.4 B-D Track Probability, P_{BD}

Separating primary from secondary tracks is one thing but being able to distinguish between tracks from the b-hadron decay and the cascade c-hadron decay is a more difficult task altogether! In both cases decay tracks are originating from a relatively high momentum state of high invariant mass which means the kinematic differences are small. In addition, the decays are often occurring close to one another in space so that it becomes difficult to assign a track with any certainty to the b-hadron decay vertex or the cascade vertex. We saw in Sect. 4.4.3 that, even with the excellent resolution available to the SLD experiment, the event-by-event efficiency to reconstruct the cascade c-hadron decay vertex was rather low. DELPHI's approach to the problem has been to aim for a more statistical separation of the cascade decay tracks by combining what small differences there are between the track types in a neural network variable, P_{BD}.

Figure 7.5 illustrates some of the variables used: Figure 7.5(a) shows distributions of the DELPHI KaonNet, described in Sect. 3.2.2, for the two different track classes. The increased chance that particles from the c-hadron vertex are a kaon, compared to the b-hadron decay, results in the distributions varying at the extreme values of the KaonNet. Figure 7.5(b) shows distributions of the track probability,

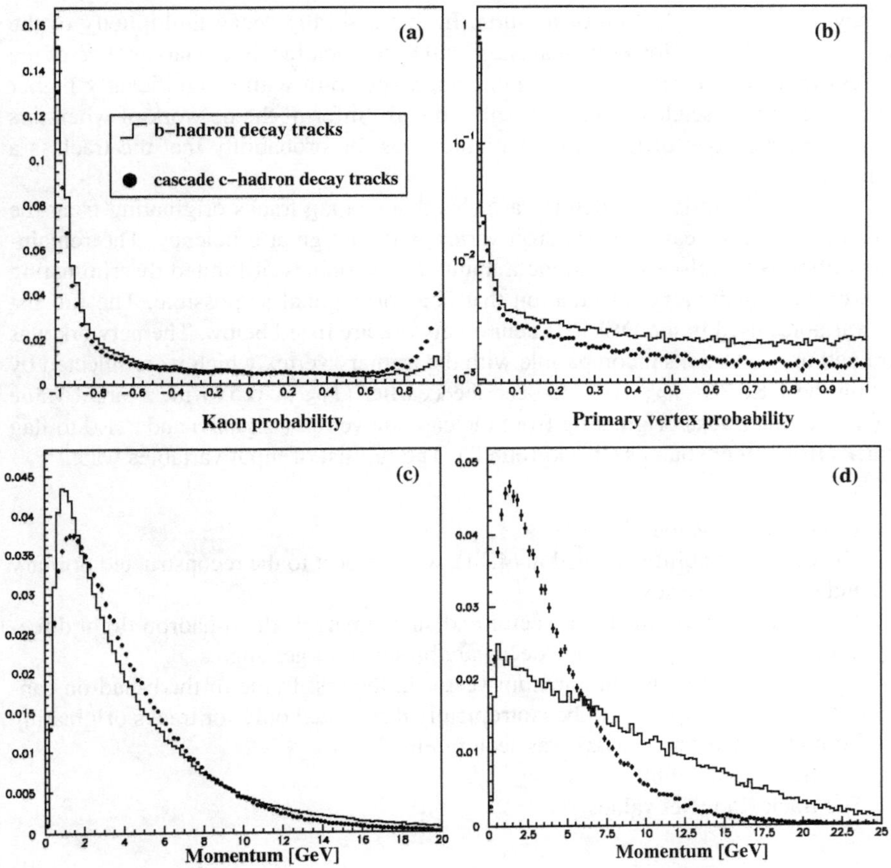

Fig. 7.5 Some inputs to the DELPHI P_{BD} neural network showing the difference between tracks from b-hadron decay (*histogram*) and tracks from the cascade c-hadron decay (*points*): (**a**) kaon probability, (**b**) primary vertex probability, (**c**) momentum of decay tracks, (**d**) momentum of decay tracks when the b-hadron decays semileptonically

defined in (4.14), with respect to the reconstructed primary vertex. Tracks from the cascade c-hadron vertex are seen to be more incompatible with originating from the primary vertex than are tracks from the b-hadron decay. Figure 7.5(c) compares distributions of track momenta for the two classes and shows that the spectrum from the c-hadron decay is slightly harder than that from the b-hadron decay even though the mean energy of the b-hadron is about 50% larger. As was the case in the discussion of particle rapidity (Sect. 7.1.2), this can be explained by the fact that the decay multiplicity of particles from the heavier b-hadrons is about two times larger than for c-hadron decays so that the available momentum is spread between many more particles. A much larger difference is seen in Fig. 7.5(d) showing the momentum distribution of tracks when the hadron has undergone a semileptonic decay, emitting an electron or a muon. In this case, the decay multiplicity of the b-hadron tends to be lower than average and each track receives a larger slice of the parent momentum, resulting in a momentum spectrum with a significantly higher mean than the cascade c-hadron distribution. To 'inform' the network of when this situation arises one of the inputs variables was the probability that the track is a lepton.

Clearly it is difficult to define variables that can tag tracks originating from the main b-hadron or cascade c-hadron vertex with any great efficiency. The remaining option is therefore to combine a number of variables of limited discriminating power but which carry information that is as orthogonal as possible. The full list of variables used in the DELPHI neural network are listed below. The network was trained only on tracks incompatible with the primary vertex which were selected by using the DELPHI P_{PS} variable described earlier ($P_{\mathrm{PS}} > 0.5$) with a target value of 1.0 to flag tracks originating from the cascade vertex ('signal') and -1.0 to flag tracks from other sources ('background'). The full list of input variables was:

- The track P_{PS} value.
- The track probability, defined in (4.14), with respect to the reconstructed primary and secondary vertex.
- The angle between the track vector and an estimate of the b-hadron flight direction. Tracks from the cascade decay are biased to larger angles.
- The angle of the track momentum vector in the rest frame of the b-hadron candidate. This is expected to be isotropically distributed only for tracks originating from the b-hadron decay as was seen in Fig. 7.3.
- The track momentum.
- The track KaonNet value.

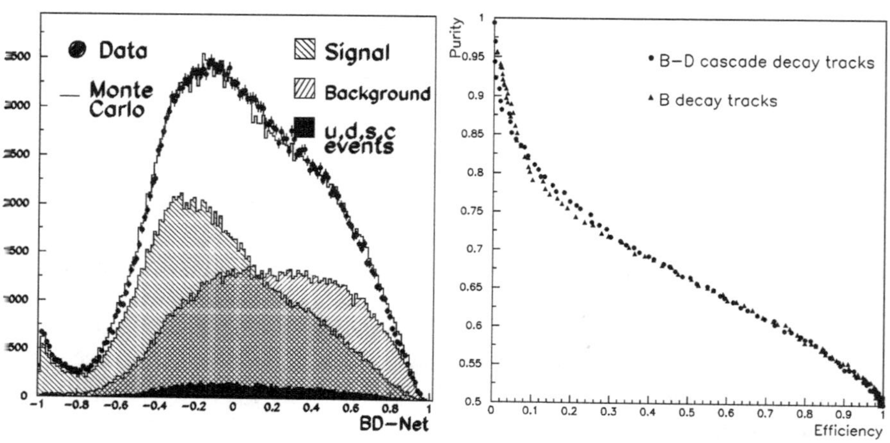

Fig. 7.6 (*left*) The DELPHI P_{BD} variable comparing data to the simulation (*right*) The purity against efficiency performance, for tagging tracks from the b-hadron or cascade c-hadron vertex, based on making progressively harder cuts on P_{BD}

- The probability that the particle is a lepton.
- Quality variables, of the kind used in the neural network P_{PS} variable.

Figure 7.6 shows the resulting output and performance of the P_{BD} network where a significant discrimination is clearly visible. This was particularly useful information for forming optimised b-hadron species tags and flavour tagging to which we now turn our attention.

7.2 Optimal b-Hadron Reconstruction

Undoubtedly the best way to know with certainty the type of b-hadron your event contains is to exclusively reconstruct it i.e. detect all of the decay products and use them to piece together the charge, mass, lifetime etc of the state. Since there are no missing decay products, the precision of the reconstructed kinematic properties of the state is only limited by the track reconstruction uncertainties in the detector.

As already noted, exclusive reconstruction never had a large role to play in LEP b-physics mainly because of the fact that each b-hadron type has a vast number of decay channels all with tiny branching fractions. The LEP $\sin 2\beta$ analyses, mentioned in Sect 5.3, provides a good example. Here exclusive reconstruction of neutral B-mesons is needed in the channel $\mathrm{B}^0 \rightarrow \mathrm{J}/\Psi\mathrm{K}^0_\mathrm{s}$ which has a branching ratio $BR(\mathrm{B}^0 \rightarrow \mathrm{J}/\Psi\mathrm{K}^0) = 8.5 \pm 0.5 \times 10^{-4}$ [7]. Applying this to the full ALEPH data set, means there are only about 30 events of this type expected *before* any reconstruction efficiencies have been applied. The analysis was therefore severely limited by the statistical precision possible.

Because of these statistical disadvantages more inclusive methods of b-hadron reconstruction dominated the LEP analyses, first through the use of 'partial' reconstruction and later through inclusive methods. We now discuss each of these two approaches.

7.2.1 *Partial Reconstruction*

The method of partial reconstruction refers to reconstructing only part of the full decay chain to give a 'hint' as to the type of b-hadron present. The advantage is that an analysis becomes sensitive to a whole class of decay channels all at once and since the whole decay chain is no longer reconstructed, the detection efficiency increases. The (small) price to pay is that event-by-event knowledge of exactly which b-hadron states are present is largely lost and has been replaced by a method that tags b-hadrons when averaged over many events. So long as the backgrounds are well understood, the huge gain in statistics from these more inclusive methods ensured that at LEP these approaches dominated.

In the charm sector the branching ratios are generally larger and the LEP experiments were able to collect samples of D^+, D^0, D^+_s and Λ^+_c with reasonable efficiency. Since a b-hadron will nearly always decay into a c-hadron, a fully reconstructed charm state provides a strong hint that a b-hadron was present. This tag becomes more powerful if in addition the charm state is matched to some other particle from the decay chain (usually a lepton) which can provide charge correlation information. Examples of this include the tagging of B^- mesons by matching a fully reconstructed D^0 with a negatively charged lepton found in the same jet from the reaction, $B^- \to D^0 \ell^- X$. Similarly, we have already seen in Sect. 4.3.1 how clean samples of B^0_d's can be obtained by matching a D^{*+} with a negatively charged lepton.

This technique was used by ALEPH [8] to provide the first evidence for the lightest b-baryon, Λ^0_b, in the decay mode $\Lambda_b \to \Lambda^+_c \ell^- \nu$. The trick was to observe the decay $\Lambda^+_c \to \Lambda X$ by detecting the decay of Λ hyperons into $p\pi^-$ and then count how many hyperons are produced with leptons of the right sign to have come from a Λ^0_b decay. Figure 7.7 (left) shows the resulting excess seen in 'right-sign' hyperon production.

Using a similar approach, but correlating reconstructed Ξ states with leptons, DELPHI obtained first evidence for a strange b-baryon state, Ξ^-_b. In 1992 DELPHI applied the technique to the decay chain, $B^0_s \to D^{(*)-}_s \ell^+ \nu_\ell X$, to gain first evidence from LEP for the B^0_s meson [9]. The presence of $D^{(*)-}_s$ mesons accompanied by oppositely charged high transverse momentum muons from the same jet, are expected to originate almost entirely from B^0_s decay and Fig. 7.7 (right) shows the hand-full of D^+_s candidates in the original sample. Note that mass measurements of b-baryons and B^0_smesons is currently dominated by by the CDF experimentsince

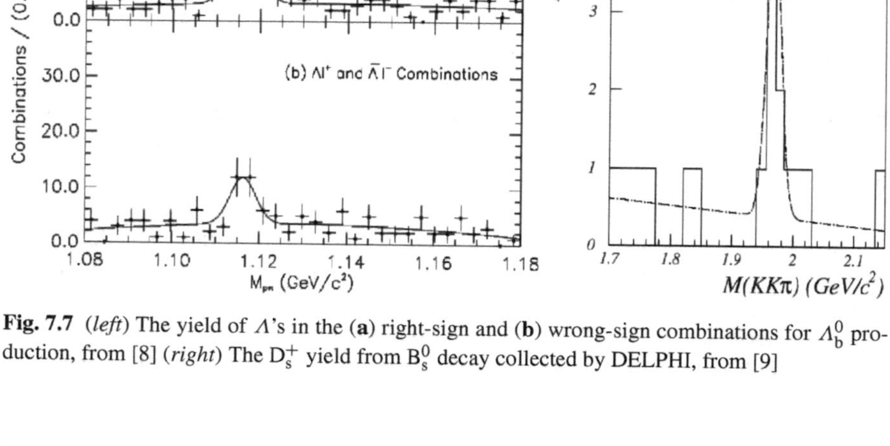

Fig. 7.7 (*left*) The yield of Λ's in the (**a**) right-sign and (**b**) wrong-sign combinations for Λ^0_b production, from [8] (*right*) The D^+_s yield from B^0_s decay collected by DELPHI, from [9]

it requires the exclusive reconstruction of the state. D-lepton (and to a lesser extent D-hadron) correlated production have been used by the LEP experiments to produce some very precise b-hadron lifetime measurements which have only been bettered in recent years by the development of fully inclusive techniques which we now discuss.

7.2.2 Inclusive Reconstruction

The later years of LEP analyses saw a burst of interest in using the inclusive tools such as secondary vertex finding, rapidity and primary-secondary probabilities. The reasons for it are clear: the huge gain in statistical power that inclusive methods bring, at some point begins to win over the systematic uncertainties introduced due to having more background in your analysis and a greater dependence on simulation. In many LEP b-physics topics the most precise, final, publications were from an inclusive approach to an analysis which began based on semileptonic decays or a specific, partially reconstructed, channel.

Of-course, all of the techniques for b-tagging described in Chap. 4 also apply here. The difference now however is that we are not only interested in just flagging the presence of a b-hadron but we want to know the type of b-hadron, its decay point in space and the kinematics. DELPHI have inclusively reconstructed B-mesons with a simple rapidity algorithm [10]: the event is divided into two hemispheres defined by the thrust axis and particle rapidities calculated with respect to this axis. In each hemisphere, the momenta of particles with $y > 1.5$ are then added together to obtain an estimate of the B-meson momentum in that hemisphere. All charged tracks are assigned the mass of a pion whereas all unassociated electromagnetic clusters are assumed to be massless photons. The resolution on the B-meson direction, was estimated to be 15 mrad for 60% of the sample and by implementing a correction based on simulation, the B-meson energy resolution was 7% for 70% of the total sample.[2] These are excellent results from such a simple algorithm and illustrate the effectiveness of inclusive methods on the LEP data. Such a rapidity algorithm could also be used as the basis for reconstructing the decay point of the B-meson by searching for a consistent secondary vertex from among the tracks with $y > 1.5$. ALEPH implemented a very similar algorithm but based on selecting particles that passed both cuts in rapidity and in impact parameter probability to define the B-candidate [11]. A resolution on the direction of the B-meson of 14 mrad was found for the higher mass candidates.

A more sophisticated approach was used by OPAL in [12]: Here, a topological method was applied to jets in order to locate possible secondary vertices. For each track in a jet containing an accepted secondary vertex a weight W was calculated to

[2] b-hadron direction and energy resolutions typically consist of a Gaussian part describing most of the data plus some broader, often non-Gaussian, component accounting for the rest.

assess the probability that the track came from the secondary vertex relative to the probability that it came from the primary vertex,

$$W = \frac{R(b/\eta)}{R(b/\eta) + R(d/\sigma)}. \tag{7.2}$$

R is a symmetric function describing the impact parameter significance with respect to a fitted vertex, b and η are the impact parameter and associated uncertainty in the $r\phi$ plane with respect to the secondary vertex, d and σ are the same quantities with respect to the primary vertex. A estimate of the b-hadron charge comes from forming the vertex charge variable summed over all tracks in the jet,

$$Q_{\text{vtx}} = \sum_{i}^{N_{\text{trk}}} W_i \cdot Q_i \tag{7.3}$$

and an estimate of the b-hadron momentum vector based on tracks follows from,

$$\boldsymbol{P}_{\text{vtx}} = \sum_{i}^{N_{\text{trk}}} W_i \cdot p_i. \tag{7.4}$$

Neutrals are also included in the momentum determination by summing all electromagnetic clusters unassociated with tracks that lie within some cone region drawn around direction $\boldsymbol{P}_{\text{vtx}}$. The vector

$$\boldsymbol{P}_B = \boldsymbol{P}_{\text{vtx}} + \boldsymbol{P}_{\text{clus}} \tag{7.5}$$

provides a first estimate of the b-hadron momentum vector. A subsequent OPAL analysis boosted the performance of this algorithm still further by combining the weight defined in (7.2) with a primary-secondary neural network probability [13].

It is inevitable that when inclusively reconstructing b-hadron states there are missing particles, due to neutrinos and detector inefficiencies, plus other effects which makes the accurate determination of the momentum of the state challenging. The estimates of momentum from all inclusive methods therefore had to be corrected for these effects, usually from simulation, and ways of doing this are discussed further in Sect. 7.2.4.

Ultimately the best performing methods of inclusive reconstruction were those based on the topological vertexing algorithms (described in Sect. 4.4.3) to determine the decay point in space and constructing some kind of weight which can be summed over in order to estimate the charge and momentum of the b-hadron. The weight should represent the probability that the particle derives from the b-hadron secondary vertex as opposed to the primary vertex and could take the form used by OPAL above, or be one of the primary-secondary probabilities reviewed in Sect. 7.1.3.

7.2.2.1 Orbitally Excited b-Hadrons

A subject where inclusively reconstructed B-mesons play a big role is in studies of the orbitally excited B^{**} mesons discussed in Sect. 1.3.6.

Most of the LEP analyses [14–17] have involved reconstructing an inclusive B^+ or B^0_d meson and then adding either a charged pion, for the case of $B^{**}_{u,d}$, or a charged kaon for studies of the B^{**}_s. Some sensitivity to the narrow states B_1 and B^*_2 is seen but no analysis managed to separate them for a mass or width measurement and instead report narrow state production fractions based on assumptions for the broad state contributions. An ALEPH analysis [18] based on exclusive $B\pi$ reconstruction had better background suppression but also failed to separate the narrow states due to a lack of statistics.

More recently there has been renewed interest in this subject triggered by a DELPHI study based on some of the high performance flavour and charge tags described here [19, 20]. Since background uncertainties were the dominant source of systematic error to the earlier inclusive measurements, this analysis concentrated on either fitting the background as much as possible from the data itself or by reducing greatly the level of background made possible by the high performance flavour and charge tags. The new analysis finds a B^{**} narrow production fraction value significantly below the previous results. A likely explanation is that the older results measured a production fraction that was mixture of narrow and broad states, a situation that could arise if the background contribution was not fully understood. Recently these hints of a B^{**} spectroscopy have been largely confirmed by the CDF and D0 collaborations who are able to study exclusive samples of $B^{**0} \to B^{(*)+}\pi^-$ decays with excellent mass resolution and high statistics. The most precise results come from the CDF analysis [21] who, have adopted some of the LEP inclusive methods in conjunction with their track-lifetime trigger and the more 'traditional' J/Ψ tag. Neural network variables combining topological, kinematic and particle identification quantities are employed to tag the presence of B^+ and a further network is trained to recognise the decay π^-. Assuming the HQET prediction for the width ratio $\Gamma(B^0_1)/\Gamma(B^{*0}_2)$, the mass difference between the B^0_1 and B^{*0}_2 states was measured to be $14.9^{+2.2}_{-2.5}(\text{stat.})^{+1.2}_{-1.4}(\text{syst.})$ MeV/c^2 and the B^{*0}_2 width was measured for the first time.

OPAL were the first LEP collaboration to study $B_s{}^{**}$ production in the channel $B^{**0} \to B^{(*)+}K^-$ and found an excess of 149 ± 31 B^+K^- pairs translating into a production fraction of, $\frac{\sigma(B^{**}_s)}{\sigma_b} = 0.020 \pm 0.006$ [12]. The new DELPHI analysis [19, 20], applying the same high performance tagging techniques as for the $B^{**}_{u,d}$ states but now to identify the partner decay K^-, isolated a signal consistent with originating from B^*_{s2} decay with a production fraction of: $\frac{\sigma(B^*_{s2})\cdot BR(B^*_{s2}\to BK)}{\sigma_b} = 0.0093 \pm 0.0020 \pm 0.0013$. Recently a CDF analysis [22] using similar inclusive neural network tags to select B^+ and K^- decay candidates, has for the first time been able to localise both narrow doublet states B^*_{s2}, B_{s1} with results that are consistent with theory and what had been measured at LEP.

The isolation of a signal due to B^{**} broad states is very challenging and has not yet been unambiguously achieved. These results however represent enormous progress in establishing the spectroscopy of the orbitally excited b-hadrons and to which the inclusive methods developed at LEP have played a lead role.

7.2.3 b-Hadron Species Tagging

There are many properties of b-events, some more obvious than others, that can be exploited to inclusively tag the presence of a particular b-hadron type. The charge of the b-hadron is clearly something which can unambiguously distinguish between e.g. a B^0-meson and a B^+ and we have already seen that variables such as vertex charge can effectively reconstruct the charge of b-hadron states. We have also already seen in Sect. 5.1 how different b-hadron states are often associated with particular particles from the fragmentation process e.g. a B^+ produced with a π^-. If they can be isolated, these associated particles provide powerful tags of specific b-hadron types. Since kaons and protons can be isolated with particle ID techniques with lower backgrounds than for pions, the most useful for this job are the kaons associated with B_s^0 production and the protons accompanying b-baryon fragmentation. In addition, the presence of B_s^0-mesons and b-baryons is signalled by kaons and protons produced as products of the weak decay of these states. Related to this observation is the number of charged pions present in a hemisphere. This is seen to be higher for B_d^0 and B^+ mesons compared to B_s^0 and b-baryons because of the higher proportion of kaons, neutrons and protons present from their production and decay processes. Another discriminator is the total energy content of hemispheres which can be somewhat lower for $B_s^0$and baryon states due to the greater number of neutrons and K_L^0 produced and which subsequently escape detection in the detector.

DELPHI has combined variables sensitive to these physics effects in a b-hadron species identification neural network [6] together with 'quality' variables such as the invariant mass of the secondary vertex and the energy of the B-candidate. The network architecture is interesting because there are four output nodes, one for each b-hadron type B^+, B_d^0, B_s^0 and Λ_b, and the network is trained to learn target *vectors* e.g. $(1, 0, 0, 0)$ for the case of a B^+ meson being present in the event hemisphere. Each output node therefore delivers the probability for the hypothesis it was trained on e.g. the first is proportional to the probability for a B^+ meson being present in the hemisphere, the second is proportional to the probability for a B_d^0 being present etc. Figure 7.8 shows distributions at each of the four output nodes (P_{B^+}, P_{B^0}, $P_{B_s^0}$, P_{Λ_b}) from an event sample that is independent from the sample used for training the network and which is about 80% pure in $Z^0 \rightarrow b\bar{b}$ events after the application of a lifetime b-tag. The level of agreement between data and simulation is remarkably good considering the complexity of the variable and it is evident that the tagging power is somewhat better for B^+ and B_d^0 compared to B_s^0 and Λ_b. This is explained by the fact that many of the discriminating variables are dependent on the separation of particles originating from the fragmentation system compared to the b-decay chain and this separation is more evident for the b-hadron species with the longer lifetimes. It is also noticeable that the tagging power at the B_d^0 node is worse than for B^+ which is mainly a consequence of the neutral states, B_d^0, B_s^0 and

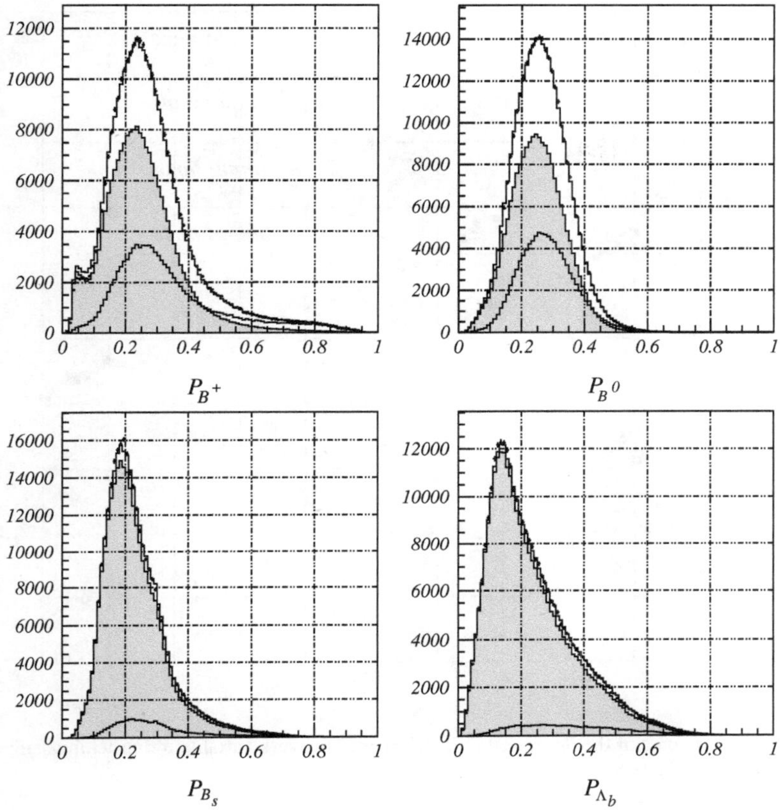

Fig. 7.8 Distributions from the four output nodes of the DELPHI b-hadron species identification neural network in the simulation (*histogram*) compared to the data (*points*). In each case the two overlaid histograms show the 'signal' component (*open histogram*) and everything else (*shaded histogram*)

Λ_b^0, all acting as sources of background to each other and so degrading the potential performance from the network.

DELPHI found that the performance of the b-hadron species identification network could be further boosted by combining the information with b-quark flavour tags which were also constructed separately for each b-hadron type and are discussed in Sect. 7.3. Note that the opposite-side hemisphere information from the identification network can not help improve the tagging performance since the fragmentation of the b and $\bar{b}$-quark are independent of each other e.g. prior knowledge of a B^+ state in one hemisphere does not predict the type of b-hadron that will appear in the opposite hemisphere. This more optimal tag was also implemented as a neural network quantity with four output nodes and trained in the same way as the initial b-hadron species identification network. The tagging performance is shown in Fig. 7.9 and represents a 5–10% improvement on the performance of the P_{B^+}, P_{B^0}, $P_{B_s^0}$, P_{Λ_b} variables alone.

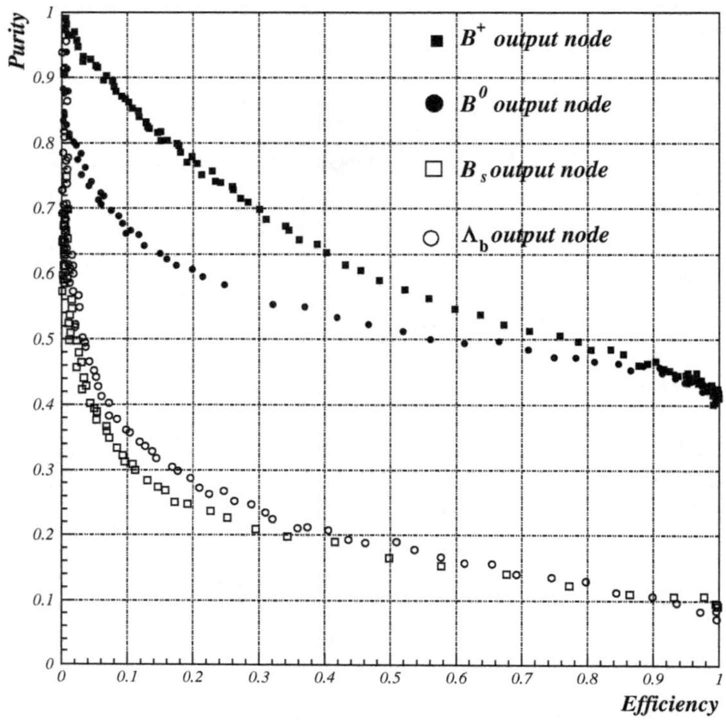

Fig. 7.9 Performance of the optimised DELPHI neural network variable for enriching samples in the various b-hadron types. From [6]

7.2.3.1 Measuring b-Hadron Lifetimes

The excellent tagging power of the DELPHI b-hadron species tag was fully utilised in a recent publication of the lifetime ratio $\tau_{\mathrm{B_u^+}}/\tau_{\mathrm{B_d^0}}$ and mean b-hadron lifetime τ_{b}. The network outputs were cut on to select samples approximately 70% pure in $\mathrm{B^+}$ and $\mathrm{B_d^0}$ mesons and a χ^2 fit made to the data by leaving the $\mathrm{B^+}$ and $\mathrm{B_d^0}$ lifetimes free in the fit but fixing the lifetimes of the $\mathrm{B_s^0}$and b-baryon components to the world average values. The result of the fit is shown in Fig. 7.10. The b-hadron species tag enabled high statistics samples to be collected at the quoted purity and resulted in the most precise measurement of the charged-to-neutral lifetime ratio from the LEP era: $\tau_{\mathrm{B_u^+}}/\tau_{\mathrm{B_d^0}} = 1.060 \pm 0.021(\mathrm{stat.}) \pm 0.024(\mathrm{syst.})$.

The current World-average lifetime measurements for b-hadrons are shown Fig. 7.11 and are seen to broadly follow the lifetime hierarchy discussed in Sect. 1.3.8. At the end of the LEP era the baryon lifetime ratio $\Lambda_{\mathrm{b}}^0/\mathrm{B_d^0}$ was uncomfortably low compared to HQET-based predictions. A recent measurement from CDF has pulled the mean lifetime for Λ_{b}^0 higher but disagrees with a similar measurement from D0 at the 3σ-level. Further precision measurements of the baryon and also of the $\mathrm{B_s^0}$ lifetime are needed before drawing conclusions about the theory.

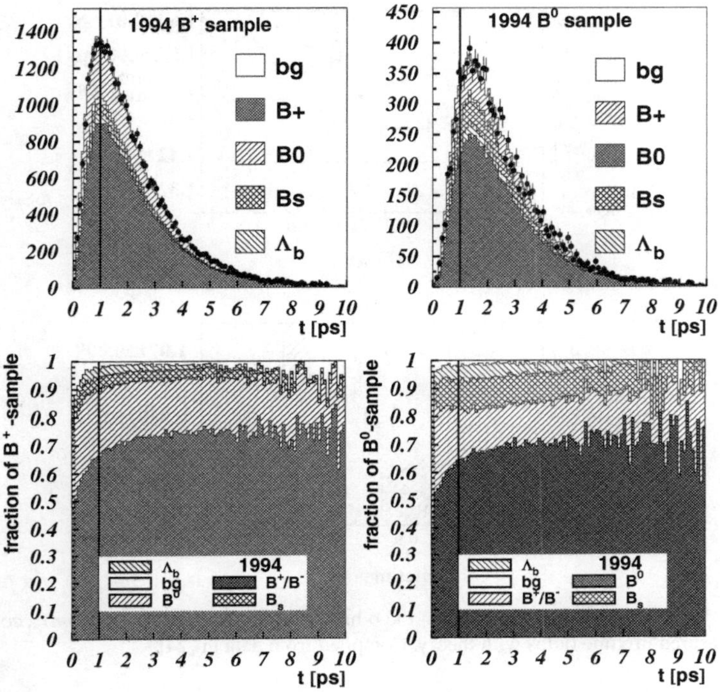

Fig. 7.10 The *upper* two plots show the result of the lifetime fit (*histogram*) to the B^+ and B^0_d-enhanced DELPHI data samples (*points*). The *lower* plots trace how the fractional composition of the samples change as a function of the lifetime. The line at 1% ps indicates an analysis cut, below which no data was used in the fit. From [23]

7.2.4 b-Hadron Energy Reconstruction

Inclusive reconstruction and partial reconstruction in semileptonic decays are subject to energy losses due to the escape of undetected neutrinos. Photons, π^0 's and neutral particles in general are also notoriously difficult to detect with any great efficiency as are tracks that disappear down cracks in the detector coverage or at low angles etc. Along with missing particles that should be there, an inclusive reconstruction is prone to including particles that do not belong there due to e.g. lack of resolution, reconstruction errors or particles originating from material interactions. Add into this the effect of giving particles the wrong mass assignments or using the assumption that all tracks are pions and all neutral clusters are photons, and it becomes clear that the b-hadron energy is a complicated quantity to accurately reconstruct.

A common method to correct the energy, which was applied for example to the estimate of the OPAL analysis [12] discussed in Sect. 7.2.2, is to consider the event as a two-body decay of the Z^0. The decay bodies are the b-hadron jet, of mass m_{jet}

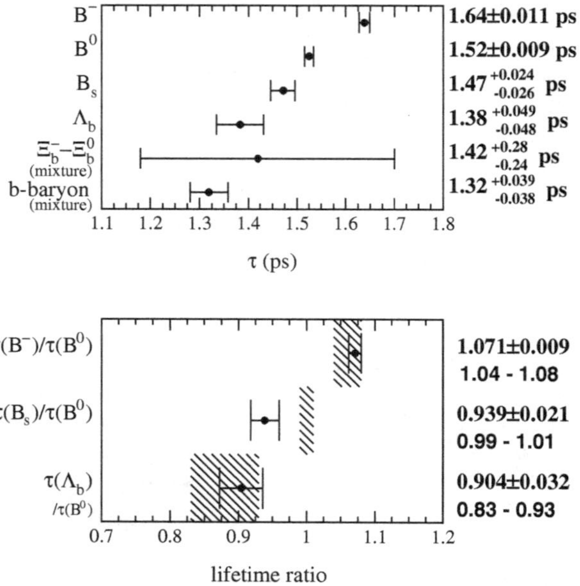

Fig. 7.11 World average measurements of the b-hadron lifetimes (*top*) and (*bottom*) a comparison of the measured lifetime ratios with theory. Compiled from data in [24]

which recoils against the whole of the rest of the event with mass m_{rest}. Demanding conservation of energy, the energy of the jet can now be expressed as [7],

$$E_{\text{jet}} = \frac{E_{\text{CM}}^2 + m_{\text{jet}}^2 - m_{\text{rest}}^2}{2E_{\text{CM}}} \tag{7.6}$$

where $E_{\text{CM}} = E_{\text{jet}} + E_{\text{rest}}$ is the event centre-of-mass energy. m_{jet} is set to the nominal B-meson mass, and m_{rest} is calculated using tracks and unassociated electromagnetic clusters from the rest of the event *outside* of the jet. The b-hadron energy is by definition,

$$E_{\text{B}} = E_{\text{jet}} - E_{\text{frag}}, \tag{7.7}$$

where E_{frag} is the fragmentation energy in the jet which is estimate from the measured quantities,

$$E_{\text{frag}} = E_{\text{jet}}^{\text{vis}} - (E_{\text{vtx}} + E_{\text{clus}})\,. \tag{7.8}$$

Here, $E_{\text{jet}}^{\text{vis}}$ is the total visible energy in the jet summed over tracks and unassociated electromagnetic clusters while, E_{vtx} and E_{clus} are the energies associated with P_{vtx} and P_{clus} from (7.5). Estimating the energy according to (7.7) gives typically a factor two or more improvement in resolution on the total energy of the b-hadron candidate compared to the initial estimate ($\boldsymbol{P}_{\text{vtx}} + \boldsymbol{P}_{\text{clus}}$). With this method, OPAL made a double-Gaussian fit to the difference between the reconstructed and the generated b-hadron energy and reported the width of the narrower Gaussian to be $\sigma = 2.8$ GeV with 95% of the entries contained within 3σ [12]. Applying this same technique in another OPAL analysis [25] led to the results in Fig. 7.12. These plots illustrate some general features which apply to a greater or lesser extent to all of the LEP b-hadron energy estimators. The energy distribution shows a large tail to lower values where the resolution attainable also starts to degrade markedly. Reasons for this include: a bias resulting from the beam energy constraint which gets worse for low energy b-hadrons; when the true b-hadron energy is low it also happens that the decay and fragmentation groups of particles in the jet are poorly separated and so the inclusive methods are not so effective; a low reconstructed energy could also be due to a poor reconstruction efficiency which would naturally tend to have worse energy resolution. Whatever the reason, these cases are best rejected by any analysis sensitive to the energy – in this case the cut was made at $E_{\text{b}} > 20$ GeV resulting in the resolution shown in Fig. 7.12(b). Although roughly centred on zero the resolution distribution shows a positive tail which can be reduced to some extent if it is required that the b-hadron has high energy or a long decay length – where the biases mentioned above

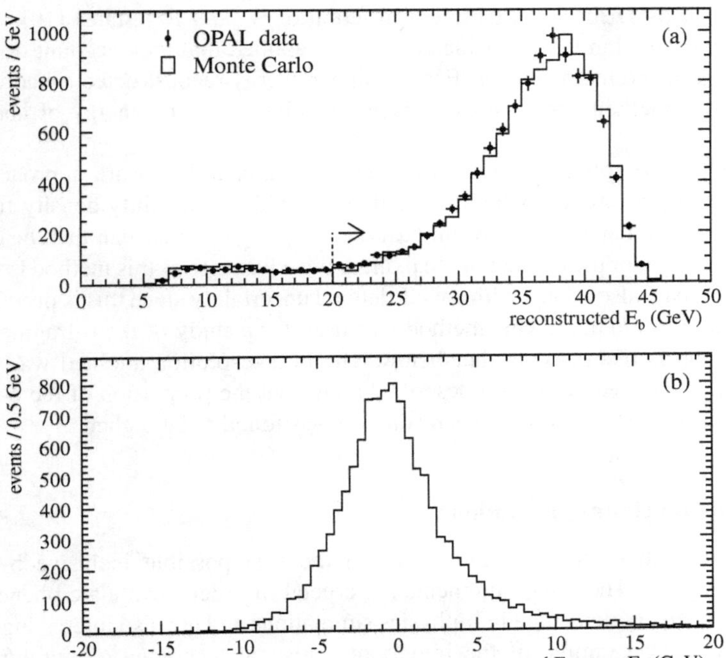

Fig. 7.12 **(a)** Corrected b-hadron energy **(b)** Resolution of the b-hadron energy in simulated b-events after a minimum energy cut at 20 GeV was applied. Results from OPAL analysis [25]

are reduced. The bulk of the resolution distribution is a core Gaussian element but there are significant non-Gaussian tails.

Use of an energy conservation constraint has also been widely applied in order to estimate the missing neutrino energy in b-hadron semileptonic decays. For example in the DELPHI analysis of $|V_{cb}|$ based on $\bar{B}^0_d \to D^{*+}\ell^-\bar{\nu}_\ell$ decays [26], (7.6) was used in terms of the same and opposite hemisphere to the B^0_d candidate decay (instead of the jet and recoil system), the missing energy in the B^0_d hemisphere is then

$$\begin{aligned} E_{\text{miss}} &= E_{\text{same}} - E^{\text{vis}}_{\text{hem}} \\ &= \frac{E^2_{\text{CM}} + m^2_{\text{same}} - m^2_{\text{opp}}}{2E_{\text{CM}}} - E^{\text{vis}}_{\text{hem}}. \end{aligned} \tag{7.9}$$

This is essentially a correction to the beam energy for events with more than two jets. DELPHI made one further correction to E_{miss} before equating it to the missing neutrino energy, which was determined from the simulation as a function of the $D^*\ell$ energy and accounted for losses due to analysis cuts, detector inefficiencies, wrong particle mass assignments etc. The final resolution on the neutrino energy was estimated to be 2.7 GeV corresponding to a 33% relative error.

Parameterising a correction to the measured energy based on the simulation was used further by DELPHI in the reconstruction of B^* and B^{**} states [10]. A four-parameter correction to the visible energy was applied that took account of neutral energy losses and inefficiencies. For the higher energy reconstructed b-hadron candidates, this method was able to achieve a resolution on the energy of about 5% (relative).

The DELPHI collaboration also implemented a neural network approach to b-hadron energy estimation which was able to unfold a probability density function for the energy on an event-by-event basis. The energy estimate can then be defined as e.g. the mean, or the median of the p.d.f. A big advantage of this method is that the energy estimate also comes with a well-defined uncertainty since this is proportional to the width of the p.d.f. This method was used for a study of the b-fragmentation function and is discussed further below. The best resolution attained was similar to the parameterised correction described above but the proportion of the b-hadron candidates reconstructed with this resolution was found to be higher.

7.2.4.1 b-Quark Fragmentation

Studies of b-quark fragmentation demand the best possible inclusive b-hadron energy estimate. These measurements are crucial in order to validate phenomenological models implemented in b-physics simulations and are also interesting experimentally as an example of the important statistical technique of *unfolding*. As introduced in Sect. 1.3.5, the non-perturbative physics of fragmentation is usually modelled by a fragmentation function whose only variable, x, is some measure of how much energy the b-hadrons receive out of the total amount available. If $f(x)$ is the phenomenological fragmentation function to determine, in practice one reconstructs a function $g(x_{\text{rec}})$ which differs from $f(x)$ due to:

(i) Finite detector resolution – events in a certain histogram bin of the true distribution can be spread into several bins of the measured distribution. This can be due to missing energy and the intrinsic energy resolution of the detector.
(ii) Limited measurement acceptance – any analysis will fail to reconstruct some class of event which will therefore only contribute to $f(x)$ and not the measured distribution $g(x_{\text{rec}})$.
(iii) Background – some fraction of events will enter an analysis by mistake which originate from other reactions and so only contribute to the measured distribution.

To be able to unfold these experimental effects and so get access to the underlying physics function is a tricky problem. Typically the unfolded solution is unstable and oscillates about the correct answer which calls for methods that involve a smoothing, or *regularisation* of the data in order to reduce statistical fluctuations. In addition, methods often rely heavily on simulation which introduces a dependence on a specific fragmentation model and tuning of that model. These are the main experimental challenges which are addressed in different ways by the experiments.

ALEPH [27] have studied b-fragmentation using $\bar{\text{B}}^0_{\text{d}} \to \text{D}^{*+}\ell^-\bar{\nu}_\ell$ decays where the reconstructed $\bar{\text{B}}^0_{\text{d}}$ energy is corrected for the missing neutrino energy using the two-body decay constraint method described above in Sect. 7.2.4. The true x-distribution is unfolded from the reconstructed distribution using the relation,

$$f_j(x) = \frac{1}{\epsilon_j} \sum_i^{N_{\text{bins}}} G_{ji} g_i(x_{\text{rec}}) \tag{7.10}$$

where the measured distribution is background-subtracted and normalised, ϵ_j is the acceptance correction in bin j and G_{ji} is a 'resolution matrix' that links b-hadrons reconstructed in bin i of the x_{rec} distribution with bin j of their truth x-value. The background, ϵ and G are all estimated from simulation. The model dependence is reduced by using an iterative approach i.e. the result of the unfolding for iteration N is based on a resolution matrix and background, weighted to agree with the result from iteration $N-1$. To avoid statistical fluctuations the distribution of the weights is smoothed with a polynomial function. The iterations continue until the change in the unfolded fragmentation function is a small fraction of the statistical errors.

The most model independent measurements however are based on the solution to the following relationship linking the measured and true distributions,

$$g(x_{\text{rec}}) = \int R(x_{\text{rec}}; x) f(x) dx + b(x). \tag{7.11}$$

Here, all resolution and acceptance effects are contained in the 'response function' R which is evaluated from simulation along with the background, $b(x)$. The method, used by OPAL [28] and DELPHI [29], employs regularised unfolding [30–32] to ensure suppression of the solution instabilities. The method is found to be almost completely independent of the initial fragmentation function used to generate the simulation sample.

Both the DELPHI and the OPAL analyses use the inclusive b-hadron techniques described in Sect. 7.2.2 – OPAL corrects the energy using the two-body decay constraint whereas the DELPHI energy estimate comes from a neural network prediction based on various reconstructed measures such as the jet energy, rapidity energy, energy in secondary vertices etc. Other measurements come from SLD [33] who use their missing mass b-tag algorithm discussed in Sect. 4.4.3 to reconstruct the b-hadron energy corrected for missing energy. This analysis uses the same method as ALEPH to unfold detector effects but without the iterative step – which makes the results somewhat model-dependent. Figure 7.13 (top) collects together the various results for b-quark fragmentation and illustrates that within the uncertainties, they are in generally good agreement for shape and peak position.

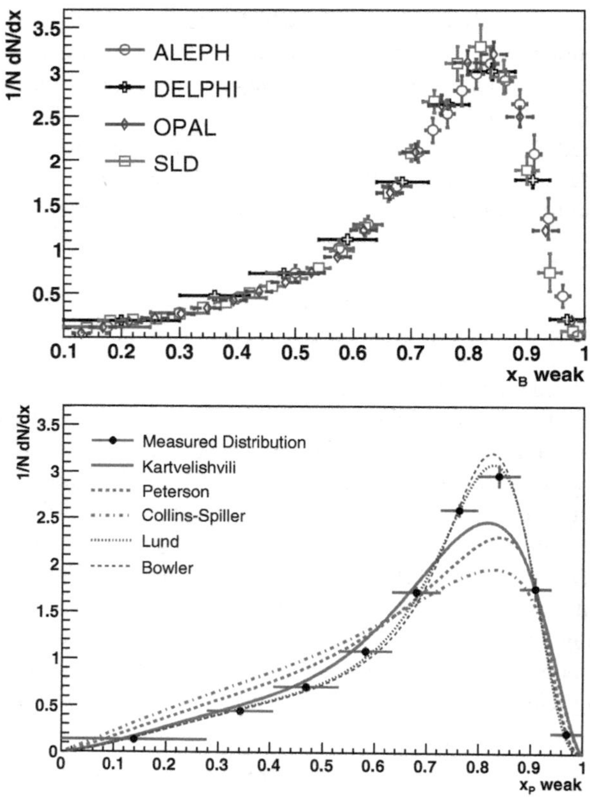

Fig. 7.13 (*top*) A comparison of all current measurements of $f(x)$ (*bottom*) DELPHI preliminary result for f(x) unfolded using the Mellin transform method (points) together with fits based on some of the most popular b-fragmentation functions. From [29]

In an attempt to draw some more universal conclusions on the fragmentation function, the DELPHI analysis goes further and uses the techniques of Mellin transforms to analytically factorise out the perturbative contribution to the fragmentation process from the non-perturbative (see Sect. 1.3.5). The result, based on the DELPHI measured x-distribution is shown in Fig. 7.13 (bottom) which illustrates that 'traditional' choices of fragmentation function, such as those from Peterson et al. or Collins-Spiller, are a poor description of b-quark data. The best fit is given by the LUND function [34]

$$f(x) = \left[\frac{1}{x}(1-x)^a \exp\left(-\frac{bm_{b\perp}^2}{x} \right) \right] \tag{7.12}$$

where $m_{b\perp}$ is the *transverse b-mass* and the free parameters are $\{a, b\}$. The same conclusion holds for the $f(x)$ distributions unfolded by other experiments and the DELPHI paper goes on to make a fit for the LUND parameters from the data which yields,

$$a = 1.48^{+0.11}_{-0.10} \; ; \; b = 0.509^{+0.024}_{-0.023}.$$

The results are model independent up to the choice made to describe the perturbative part of $f(x)$. The above result was based on PYTHIA 6.156 parton shower but a NLL QCD calculation could equally well be used. This result therefore represents a measurement of the b-fragmentation function which should be valid in environments other than LEP with the caveat that to get the same results, the same description of the perturbative part of $f(x)$ must be used.

7.3 Optimised Flavour Tagging

The DELPHI collaboration made extensive studies into how flavour tagging can be approached in the most optimal way. The conclusion of this work formed part of the BSAURUS package [6] consisting of a collection of inclusive flavour/B-species tags which found application in a diverse range of analyses including: lifetime ratio $\tau(B^+)/\tau(B^0)$ [23], A_{FB}^{b} [35], b-quark fragmentation function $f(z)$ [36], studies of orbitally excited B^{**} mesons [10], the wrong-sign charm rate in $Z^0 \rightarrow b\bar{b}$ events [37] and B_s^0 oscillations [38]. The main conclusions of the BSAURUS studies of inclusive b-physics are:

(1) Potentially the best performance comes from combining variables, $X = \{x_1, x_2, \ldots, x_n\}$, with a neural network which can optimally make use of the cross correlations between inputs. These inputs should contain discriminating variables able to distinguish between a b and $\bar{b}$-quark and also quality variables that provide information as to how reliable the inputs might be on an event-by-event basis.

(2) The problem should be addressed separately for each b-hadron type. This is important because, as Fig. 5.1 shows, the expected fragmentation track charge correlations with the parent b-quark depend on which species of meson we are tagging. There are experimental issues too that impose differences in tagging power between the b-hadron species: e.g. use of the vertex charge variable can only tag b-quark flavour for the case of charged b-hadron states and the charge correlations we use are carried by different types of particles e.g. pions or kaons, which will naturally have different detection efficiencies. These kind of effects mean that the different b-hadron types are not expected to have the same tagging efficiency and purity performance and so a more optimal approach would be to treat them separately. For the neural network tag, if the b-hadron types were not treated separately, we would be relying on the training phase of the network to 'learn' the many correlations between each variable and a particular b-hadron type. Although possible, this kind of complex task is best avoided in order to make finding the global minimum as quick and reliable as possible when training the neural networks.

(3) The network should be trained on a track-by-track basis to return the *conditional probability* $P(\text{same } Q|X_i)$. This is the probability for a track i to have the same sign of charge as the b-quark in the b-hadron, given a vector of measurable quantities X_i associated with the track. The track-level probabilities can then be combined in a likelihood ratio to provide a flavour tag at the level of the event-hemisphere,

$$F(hem.)_k = \sum_i^{\text{N-tracks}} \ln\left(\frac{1 + P_k(\text{same } Q|X_i)}{1 - P_k(\text{same } Q|X_i)}\right) \cdot Q_i \tag{7.13}$$

for $k = \text{B}^+, \text{B}^0_\text{d}, \text{B}^0_\text{s}$ or b-baryon and Q_i is the track charge.

(4) The whole process of constructing these flavour tags can be repeated for the task of tagging the production flavour and the decay flavour separately. In the absence of $\text{B}^0 - \bar{\text{B}}^0$ mixing the b-quark type is the same in both cases but for inclusive analyses of b-mixing, knowledge of the flavour at production and decay time is essential. In addition to the hemisphere flavour tags described above, making use of the 'standard' flavour tags such as jet charge (for production flavour) and vertex charge (for decay flavour) can help the final performance.

Input variables to the DELPHI neural networks designed to return the probabilities $P_k(\text{same } Q|X_i)$ included:

- *Particle identification variables:* kaon, proton, electron and muon classification quantities.
- *Primary-B-D vertex separation variables:* the probability P_{PS}; the probability P_{BD}; $\frac{P_{\text{BD}} - P_{\text{BD}}^{\text{min}}}{\Delta BD}$, where $P_{\text{BD}}^{\text{min}}$ is the minimum P_{BD} value out of all tracks in the hemisphere above a P_{PS} value of 0.5 and ΔBD is the difference between

$P_{\rm BD}^{\rm max}$ and $P_{\rm BD}^{\rm min}$; the momentum of the track in the B-candidate centre of mass frame; the helicity angle of the track vector in the B-candidate centre of mass frame.
- *Track-level quality variables:* a measure track quality as used in the definition of $P_{\rm PS}$ and the track energy.
- *Hemisphere-level quality variables:* the gap in rapidity between the track of highest rapidity below a $P_{\rm PS}$ cut at 0.5 and that of smallest rapidity above the cut at 0.5; the number of tracks in the hemisphere (a) passing standard quality cuts and (b) with a $P_{\rm PS}$ value greater than 0.5; the mass of a b-hadron candidate secondary vertex fitted in the hemisphere; the χ^2 probability of the secondary vertex fit; the energy of the b-hadron candidate; the error on the vertex charge measurement as defined in (5.4).

For the task of reconstructing the decay flavour, all of the above variables were used, whereas for production flavour tagging of $\rm B_d^0$-mesons the variables related to decay properties (B-D vertex separation and lepton charge) can be dropped. The neural networks were trained on target values of $-1\,(+1)$ if the track charge is the same sign (opposite sign) to the charge of the b-quark. Distributions of $P_k(\text{same } Q|X_i)$ where the network training was based on the production flavour are shown in Fig. 7.14 for all four types of b-hadron species. Note that these plots are made for fragmentation track candidates only, selected by cutting on the primary-secondary track probability $P_{\rm PS} < 0.5$. The results are dominated in each case by the charge correlation between the leading fragmentation track and the hadron b-quark charge. As seen in Fig 5.1 this correlation flips sign between the neutral and charged B-mesons. For b-baryons, the charge correlation comes from the associated production of a proton which carries the same sign of correlation as for the neutral B-mesons. In fact, the b-baryon flavour probability provided the best correlation to production flavour when compared to the other b-hadrons and was primarily because of the excellent performance of the DELPHI proton identification (see Sect. 3.2.2). Similarly, $\rm B_s^0$ flavour tagging benefited from the excellent charged-kaon identification (see Sect. 3.2.2). The observation that the $\rm B^+$ tag performance was somewhat better than that for $\rm B_d^0$ is expected from the simple fragmentation picture of Fig. 5.1 which shows that charged B-mesons may be produced with a charged pion or kaon whereas $\rm B_d^0$'s are only produced with a charged pion. Furthermore, if a neutral $\rm \bar{K}^{0*}$ is produced in association with the $\rm B_d^0$, the charged kaon from the subsequent decay has the 'wrong' charge correlation.

A flavour tag at the hemisphere level was constructed using (7.13) where the sum was over all tracks in the hemisphere with $P_{\rm PS} < 0.5$ for the production tag and all tracks with $P_{\rm PS} > 0.5$ for the decay flavour tag. DELPHI used these hemisphere level quantities as the basis for flavour tags tailored to the specific analysis task. In general however, they formed inputs to further neural network variables based on constructions of the form,

$$I_{\rm k} = \left(F(\text{hem.})_k^{\text{dec.}} - F(\text{hem.})_k^{\text{frag.}} \right) \cdot P_k, \tag{7.14}$$

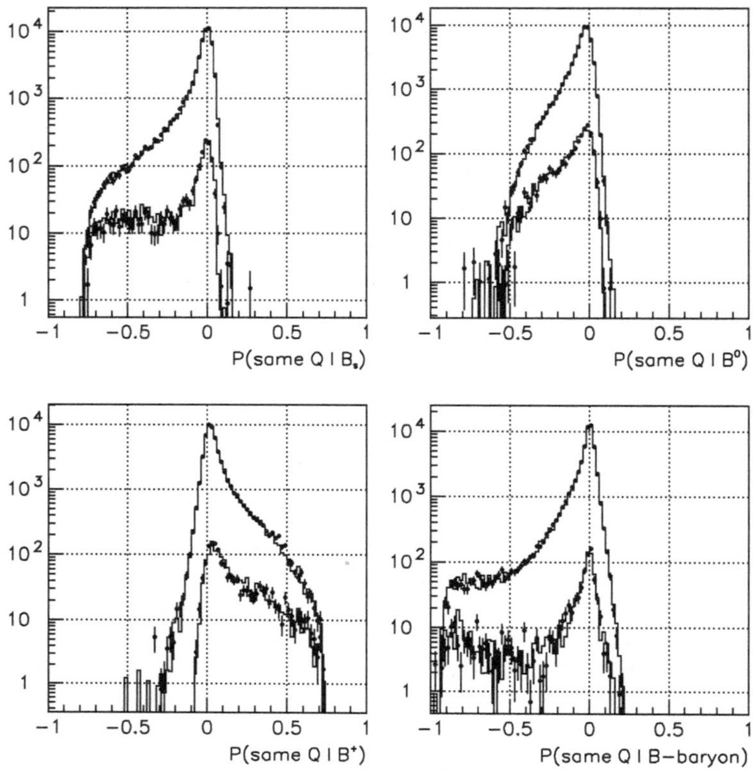

Fig. 7.14 The track-level conditional production flavour probabilities for all B-hadron types, comparing DELPHI data (*points*) and simulation (*histogram*). The plots with larger normalisation correspond to the normal mixture of B-hadron types whilst the distributions with smaller normalisation are for samples enhanced in that B-type. From [6]

where P_k are the DELPHI B-species identification tags that were described in Sect. 7.2.3. This type of construct brings information to the neural network about when the fragmentation and decay systems agree on the b-quark charge present ($I \sim 0$), when only one makes a strong prediction (I is large and either positive or negative) plus all other combinations somewhere in between these two extremes. The network would be trained to recognise b-hadron production or decay flavour at the hemisphere level and is able learn during the training phase the correlations between I and the target value and adjust its internal weights accordingly. See [6] for further details.[3]

[3] Note that B_s^0 mesons oscillate many times during an average lifetime and so in this case, the decay flavour tag has no correlation with the production flavour and would be removed from the input definition i.e. $I_{B_s} = F(\text{hem.})^{\text{frag.}}_{B_s} \cdot P_{B_s}$.

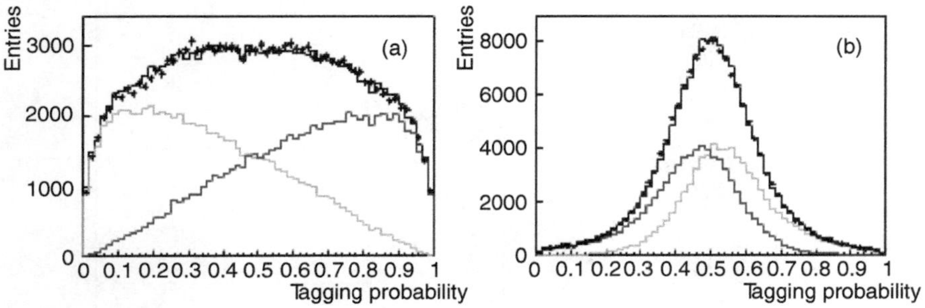

Fig. 7.15 The output of the BSAURUS (**a**) production and (**b**) decay flavour tags as used in the DELPHI B_s^0 oscillations analysis [38]. The distributions for hadrons containing a b and $\bar{b}$ quark in the simulation are indicated in light and dark grey. The data are the points

The DELPHI A_{FB}^b [35] and B_s^0 oscillations [38] analyses were based on the BSAURUS production flavour neural network tag. The network output from the B_s^0 analysis is displayed in Fig. 7.15(a), comparing data and simulation in an event sample enhanced in b-events, and showing the contributions of b-hadrons containing a b or $\bar{b}$ quark in the simulation. Neutral B-meson oscillation analyses are a particular challenge for flavour tags since knowledge of both the production and decay flavour are required. The B_s^0 analysis used the BSAURUS decay flavour tag trained specifically on B_s^0 mesons – see Fig. 7.15(b). The purity for a correct tag reaches about 62% for 100% tagging efficiency and is shown in Fig. 7.16 along with the performance of all the BSAURUS decay flavour tags.[4] Note that the main reason why the decay tag works best for B^+ mesons is related to the relatively long lifetime. Many of the network input variables are more powerful for a cleaner separation of fragmentation and decay tracks – this is also evident when constructing B-species tags (see Sect. 7.2.3). As was noted previously for the production flavour tag, the better performance for Λ_b compared to B_s^0 can be explained by the excellent proton tagging of DELPHI which provide relatively pure tag of b-baryons.

The flavour separation power of the BSAURUS flavour tags in b-quark hemispheres is quantified and compared in Fig. 7.17 which shows the tag purity against fraction of hemispheres classified. The production flavour tag (*closed dots*) for example, is found to give a purity of correct tags of over ∼70% for a 100% tagging efficiency which represents a ∼10% absolute improvement in purity over a tag based on jet-charge alone.

The development of such inclusive flavour tags were instrumental in allowing the LEP experiments to set ever more aggressive lower limits on the B_s^0 oscillation frequency e.g. the DELPHI analysis concluded that $\Delta m_s > 5.0$ ps at 95% CL. Although the first measurement of B_s^0 oscillations had to wait for the large sample

[4] Since the flavour tags are symmetric distributions about zero, these plots are obtained by cutting in a symmetric band around zero to larger and larger tag values. The purity is then the fraction of all hemispheres passing the cut that are correctly tagged and the x-axis is the fraction of correct tags out of the total possible.

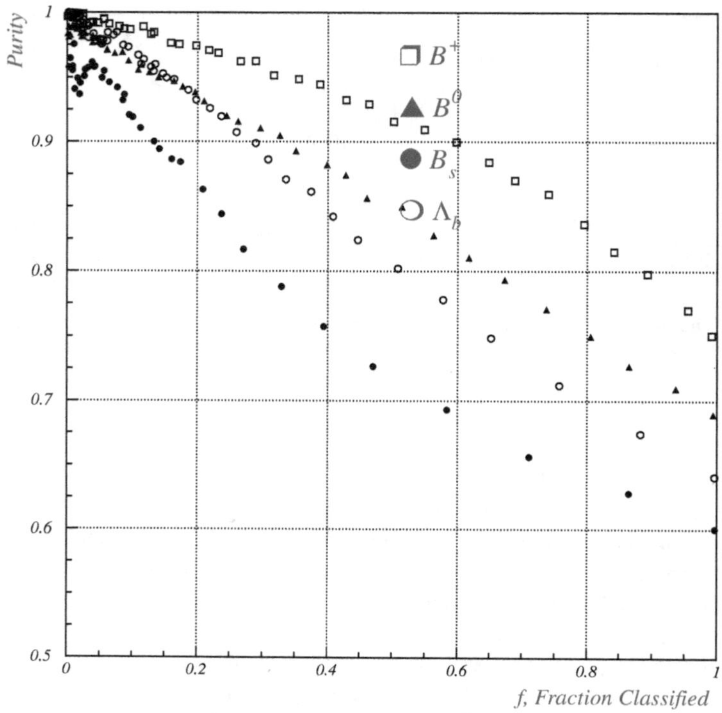

Fig. 7.16 The BSAURUS hemisphere decay flavour tag performance for all b-hadron types. From [6]

of exclusively reconstructed B_s^0 decays from Run II of the Tevatron, BSAURUS production flavour methods played a role in the CDF analysis and is discussed again in Chap. 8.

Finally we note that there is extra information in the flavour tag of the opposite side hemisphere F(opp.hem) since the b and $\bar{b}$ quark tend to fly off back-to-back. For an analysis like A_{FB}^{b}, which is highly sensitive to the angular correlation between the quark pair, only information from the same side hemisphere should be used in the flavour tags for risk of introducing systematic angular effects. However where the best possible raw flavour tagging power is needed, the opposite side information can be incorporated by modifying (7.14) as follows

$$I_k = F(\text{hem.})_k^{\text{dec.}} \cdot \left(F(\text{opp. hem.}) - F(\text{hem.})_k^{\text{frag.}} \right) \cdot P_k. \tag{7.15}$$

Figure 7.17 shows that including opposite side flavour tag information into the BSAURUS production (*open dots*) and decay flavour tagging (*open squares*) brought significant performance gains and tagging purities of almost 80% at 100% efficiency were registered [6].

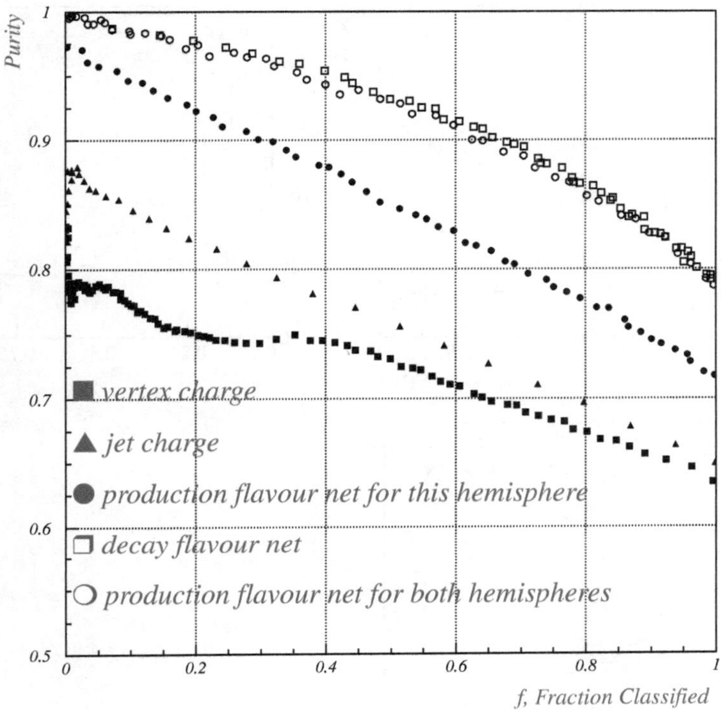

Fig. 7.17 The performance of the DELPHI optimised production and decay-flavour tags compared to jet charge and vertex charge. From [6]

7.3.1 Charm Counting

We saw in Sect. 1.3.8.2 that the mean number of charm quarks and anti-quarks produced per b-decay (n_c) is of interest because of the (anti)-correlation with the B semi-leptonic branching ratio which results in a stringent test of HQET. Equation (1.20) linked n_c to the decay branching ratios of b-quarks to final states containing no-charm and two-charm quarks.

The high performance double hemisphere decay flavour tag described above was utilised in a DELPHI analysis [37] that was sensitive to these double charm decays of the b-quark by identifying 'wrong-sign' charm states produced at the W vertex, $W^- \rightarrow \bar{c}s$ in $b \rightarrow cW^-$ events. Charmed mesons in $Z^0 \rightarrow b\bar{b}$ events were exclusively reconstructed in the modes, $D^0 \rightarrow K^-\pi^+$, $D^+ \rightarrow K^-\pi^+\pi^+$and $D_s^+ \rightarrow \phi(1020)\pi^+ \rightarrow K^+K^-\pi^+$and the decay flavour tag used to separate the wrong- and right-sign contributions. Figure 7.18 illustrates the extremely clean separation achieved which enabled the wrong-sign contribution to be fitted. The analysis found that b-hadrons decay to wrong-sign charm mesons about 20% of the time with half of this rate coming from D_s^+ production and the other half from D^+ and D^0 production.

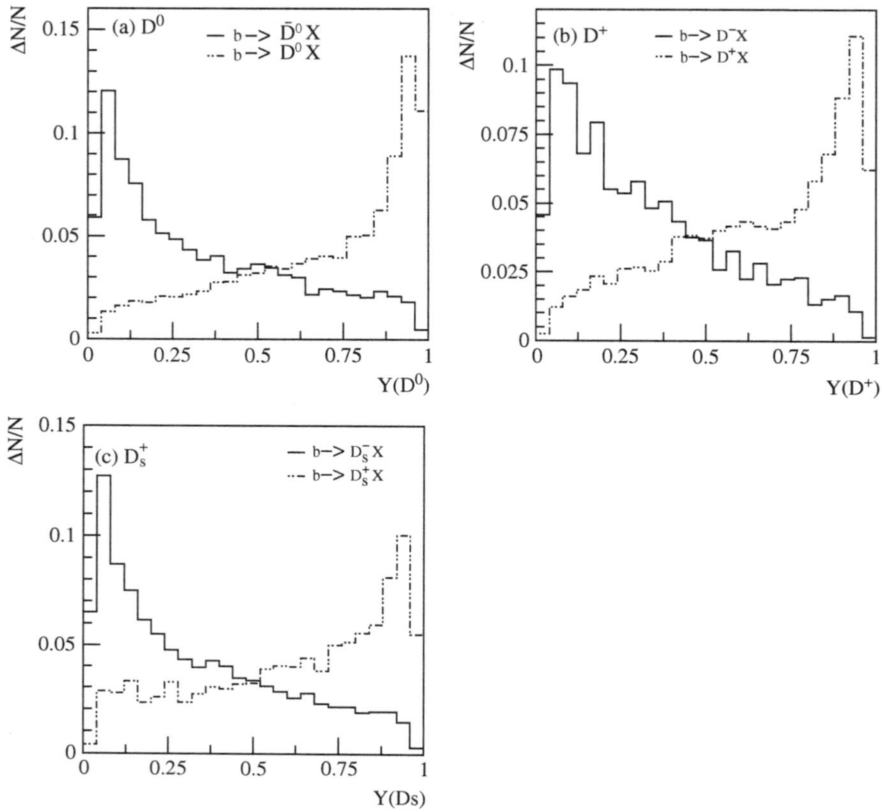

Fig. 7.18 The separation of right- and wrong-sign b-hadron decays to D^0, D^+ and D_s^+ given by the DELPHI decay flavour tag on simulated events. From [37]

Other results on the rate of no-charm and two-charm production in b-hadron decays has come from DELPHI [39] and ALEPH [40] in addition to experiments running at the $\Upsilon(4s)$. Figure. 7.19 represents the combination of this collection of measurements from [41], translated into measurements of n_c and taking into account correlated systematic errors and a common set of model assumptions for missing baryon and charmonium states. The $BR(\mathrm{b} \to \ell^- \bar{\nu}_\ell \mathrm{X})$ value from Z^0 data in Fig. 7.19 has been averaged in a similar way but in addition, a correction factor has been applied to account for the different mixture of b-hadrons in the Z^0 data compared to the mix present at the $\Upsilon(4s)$.

Over the years, the $BR(\mathrm{b} \to \ell^- \bar{\nu}_\ell \mathrm{X})$ vs n_c plot has given cause for concern either because the high energy and low energy results do not agree or the measured points fall outside of the theoretical 'comfort zone', or both. Currently it is clear that the measurements from the Z^0 and the $\Upsilon(4s)$ are compatible within the measurement uncertainties. A lingering worry however is that the data from the $\Upsilon(4s)$ is only described by theory with QCD corrections evaluated at scales (μ) that are at

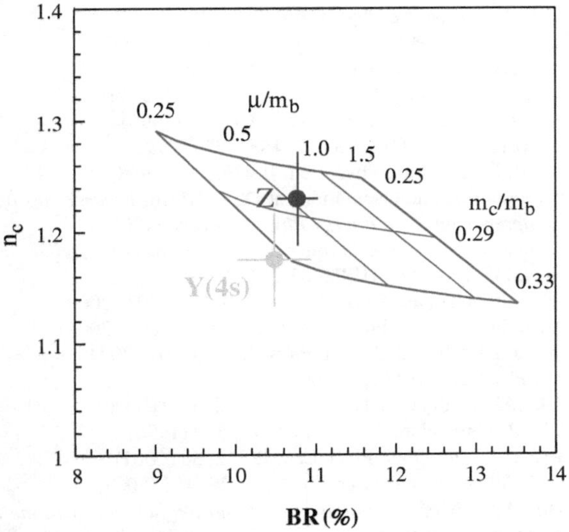

Fig. 7.19 Comparing measurements in the n_c vs $BR(\mathrm{b} \rightarrow \ell^-\bar{\nu}_\ell \mathrm{X})$plane from the $Z^0$and the $\Upsilon(4s)$, from [41]. The semileptonic branching fraction at the $\Upsilon(4s)$ comes from CLEO [42] and the theoretical prediction zone is from [43]

the extreme low end of what is reasonably allowed and for c-quark to b-quark mass ratios at the extreme high end. Better experimental precision would help clarify the situation but also the large spread in the theory prediction for $BR(\mathrm{b} \rightarrow \ell^-\bar{\nu}_\ell \mathrm{X})$ as a function of μ together with the low value favoured for this parameter, could be indicating that calculations of QCD corrections to higher orders are required.

References

1. C. Peterson: Pattern recognition in high energy physics with neural networks. In: *Proceedings, Erice91, QCD at 200 TeV, Ettore Majorana Int. Sci. Ser., Physical sci.*, vol 60, L. Cifarelli, Y. Dokshitzer eds. Plenum, New York (1992) pp. 149–163
2. OPAL Collab., K. Ackerstaff et al.: Zeit. Phys. C **73**, 397 (1997)
3. OPAL Collab., G. Abbiendi et al.: Eur. Phys. J. C **8**, 217 (1999)
4. OPAL Collab.,G. Abbiendi et al.: Phys. Lett. B **482**, 15 (2000)
5. ALEPH Collab., R. Barate et al.: Phys. Lett. B **492**, 259 (2000)
6. Z. Albrecht, T. Allmendinger, G. Barker, M. Feindt, C. Haag, M. Moch: BSAURUS-A Package For Inclusive B-Reconstruction in DELPHI., hep-ex/0102001 (2001)
7. K. Hagiwara et al.: Phys. Rev. D **66**, 010001 (2002) and 2003 off-year partial update for the 2004 edition available on the PDG WWW pages (URL: http://pdg.lbl.gov/)
8. ALEPH Collab., D. Buskulic et al.: Phys. Lett. B **278**, 209 (1992)
9. DELPHI Collab., P. Abreu et al.: Phys. Lett. B **289**, 199 (1992)
10. Z. Albrecht, G.J. Barker, M. Feindt, U. Kerzel, M. Moch: A Study of Excited b-hadron States with the DELPHI Detector at LEP, contribution to EPS 2003, Aachen, DELPHI 2003-029-CONF-649
11. ALEPH Collab., D. Buskulic et al.: Z. Phys. C **69**, 393 (1996)

12. OPAL Collab., R. Akers et al.: Z Phys. C **66**, 19 (1995)
13. OPAL Collab., G. Abbiendi et al.: Eur. Phys. J. C **23**, 437 (2002)
14. OPAL Collab., R. Akers et al.: Z. Phys. C **66**, 19 (1995)
15. DELPHI Collab., P. Abreu et al.: Phys. Lett. B **345**, 598 (1995)
16. ALEPH Collab., D. Buskulic et al.: Z. Phys. C **69**, 393 (1996)
17. L3 Collab., M. Acciarri et al.: Phys. Lett. B **465**, 323 (1999)
18. ALEPH Collab., R. Barate et al.: Phys. Lett. B **425**, 215 (1998)
19. G.J. Barker: B Production and Oscillations at DELPHI. In: *Proceedings of the International Europhysics Conference on High Energy Physics*, Aachen (2003)
20. M. Moch: In: *Proceedings of the EPS International Europhysics Conference on High Energy Physics*, Lisbon, 2005, Proc. Sci. HEP2005 232 (2006)
21. CDF Collab., T. Aaltonen et al.: Phys. Rev. Lett. **102**, 102003 (2009)
22. CDF Collab., T. Aaltonen et al.: Phys. Rev. Lett. **100**, 082001 (2008)
23. DELPHI Collab., J. Abdallah et al.: Eur. Phys. J. C **33**, 307 (2004)
24. K. Amsler et al.: Phys. Lett. B **667**, 1 (2008)
25. OPAL Collab., G. Abbiendi et al.: Eur. Phys. J. C **12**, 609 (2000)
26. DELPHI Collab., P. Abreu et al.: Zeit Phys. C **71**, 539 (1996)
27. ALEPH Collab., A. Heister et al.: Phys. Lett. B **512**, 30 (2001)
28. OPAL Collab., G. Abbiendi et al.: Eur. Phys. J. C **29**, 463 (2003)
29. DELPHI Collab.: *A Study of the b-quark fragmentation function with the DELPHI detector at LEP I and an averaged distribution at the* Z^0 *pole*, to be submitted to Eur. Phys. J
30. V. Blobel: *The RUN manual: Regularized Unfolding for High-Energy Physics*, OPAL Technical Note TN361 (1996)
31. V. Blobel: *Unfolding Methods In High-Energy Physics Experiments*, DESY 84-118 (1984)
32. V. Blobel: In *Proceedings of the 1984 CERN School of Computing*, CERN 85-02 (1985)
33. K. Abe et al.: Phys. Rev. D **65**, 092006 (2002)
34. B. Andersson, G. Gustafson, B. Soderberg: Z. Phys. C **20**, 317 (1983)
35. DELPHI Collab., J. Abdallah et al.: Eur. Phys. J. C **40**, 1 (2005)
36. G.J. Barker, M. Feindt, U. Kerzel, L. Ramler: *A Study of the b-Quark Fragmentation Function with the DELPHI Detector at LEP 1*, contribution to ICHEP 2002, Amsterdam, DELPHI 2002-069-CONF-603
37. DELPHI Collab., J. Abdallah et al.: Phys. Lett. B **561**, 26 (2003)
38. DELPHI Collab., J. Abdallah et al.: Eur. Phys. J. C **28**, 155 (2003)
39. DELPHI Collab., P. Abreu et al.: Phys. Lett. B **426**, 193 (1998)
40. ALEPH Collab., R. Barate et al.: Eur. Phys. J. C **4**, 387 (1998)
41. ALEPH, CDF, DELPHI, L3, OPAL, SLD: Combined results on b-hadron production
42. CLEO Collab., B.C. Barish et al.: Phys. rev. Lett. **76**, 1570 (1996)
43. M. Neubert, C.T. Sachradja: Nucl. Phys. B **483**, 339 (1997)

Chapter 8
Conclusion and Next Steps

We have seen that from a starting point where essentially no dedicated b-physics program was envisaged, the b-physics legacy of the LEP era is a diverse collection of results which in many cases brought the subject into the realms of precision physics. Two key developments above all others made this possible: the first was the construction of precision vertex detectors based on silicon strip devices and the second was the widespread application of inclusive b-physics analysis techniques.

In Chap. 2 we discussed how, within a few years of the machine start-up, all the LEP experiments were taking data with multi-layer silicon vertex detectors providing impact parameter resolutions of $< 30\ \mu\text{m}$ in 3-dimensions and each equipped with advanced tracking and particle identification systems. We saw in Chap. 3 how progress in understanding detector alignment, calibration, tracking and particle identification were an essential precursor to reliable, precision, physics measurements. As for the analyses themselves, Chaps. 4–6 traced the development of key techniques in tagging $Z^0 \to b\bar{b}$ events, tagging b-quark charge and the double-hemisphere methods that reduced our reliance on uncertain models implemented through simulated data sets. For many of the LEP 'final' b-physics publications, the incessant drive to higher and higher precision, also meant use of inclusive analysis methods. These culminated in some very sophisticated algorithms for b-hadron species tagging and b-quark flavour tagging (described in Chap. 7), which attempt to use every scrap of available information to fully exploit any useful correlations.

These innovations were so effective that in some areas of b-physics the precision was such that deviations from Standard Model predictions could be tested. Notable examples of this include the R_b crisis discussed in Sect. 4.6.1 and the forward-backward asymmetry of b-quark production which emerged as the single most sensitive test of the Standard Model at LEP. It is perhaps ironic therefore that at the end of the LEP era, A_{FB}^{b} has emerged as the one electroweak parameter showing a significant discrepancy with the Standard Model expectation as discussed in Sect 6.4! These tests of the Standard Model rely on measurements at the few-percent level and it is testament to the invention, skill, planning and shear determination of the LEP collaboration experimentalists that this position was reached.

So how do some of these ideas carry over into other environments? LEP phase I running at the Z^0 resonance was followed by a period of running at higher C.O.M.

G.J. Barker, *b-Quark Physics with the LEP Collider*, STMP 236, 163–168,
DOI 10.1007/978-3-642-05279-8_8,

energies of between 130 and 208 GeV. This LEP II phase was never going to produce the statistical precision for b-physics achieved in the first phase since: (a) event rates were more than two orders of magnitude lower than at LEP I and, (b) there are no b-quarks produced in $W^{\pm}$ decay since $m_t >> m_W$. It was none-the-less important to check the Standard Model prediction for the evolution of the electroweak variables R_b and A_{FB}^b at energies above the Z^0 pole. All the LEP collaborations have now presented studies of $b\bar{b}$ production in LEP II data [1] and find consistency with the Standard Model prediction that R_b falls and A_{FB}^b rises slowly at C.O.M. energies 100 GeV or so above the Z^0 mass. Furthermore, b-quark tagging had a crucial role to play in searches for the Higgs boson since Higgs decay to $b\bar{b}$ is the dominant decay mode for masses of the Higgs up to the WW threshold. With the low event rates, it was important to develop b-taggers with the highest possible efficiency and for this the LEP I inclusive approach of optimally combining a host of sensitive variables generally gave the best performance. As reported in Sect. 4.5, this 'kitchen-sink' approach was able to reach tagging efficiencies of around 20% with essentially no background and was undoubtedly the main factor behind the aggressive limit on the Higgs mass achieved at the end of LEP II running: $m_H > 114$ GeV/c^2 [2].

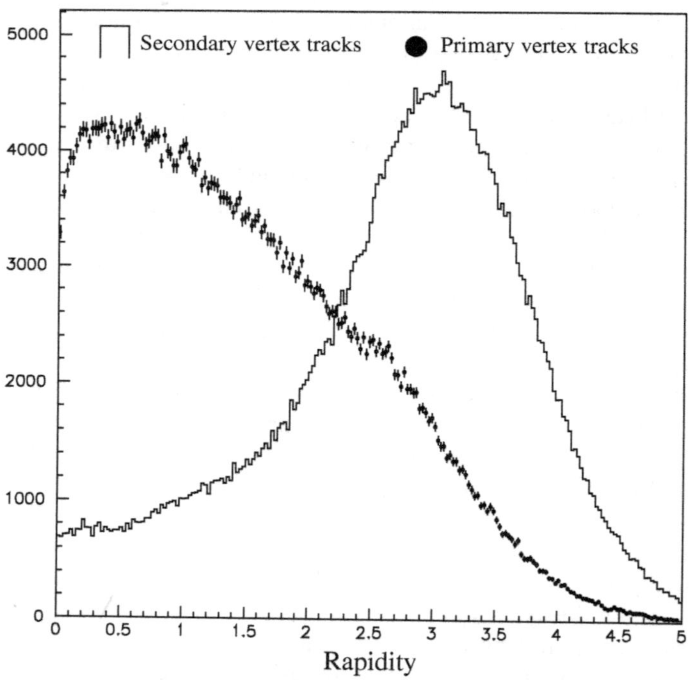

Fig. 8.1 Comparing the rapidity distributions of fragmentation tracks originating from the primary vertex, to tracks originating from some displaced secondary in the LEP II data set at energies above the Z^0-pole

To achieve these results many of the tools used to analyse the LEP I data needed to be revised to work in the new environment. For example, some of the neural network variables of the DELPHI inclusive b-physics program described in Chap. 7 had to be modified and re-trained to be relevant at the higher energies: $Z^0 \rightarrow b\bar{b}$ event samples isolated from the LEP II data were naturally more 'jetty' than at the Z^0 pole and the jet clustering variable d_{cut} (described in Sect. 4.4.1) needed to be double the LEP I value in order to reproduce the same 3-jet to 2-jet rate seen in the LEP I data. The cuts used on the rapidity variable, which was so effective at LEP I to isolate B-meson decay tracks, also needed to change. Figure 8.1 presents distributions of track rapidity originating from primary and secondary vertices and should be directly compared to Fig. 7.1 showing the same thing at LEP I. The optimal separation cut, which was around $y = 1.6$ at LEP I, increased to somewhere between $y = [2, 2.5]$ at LEP II mainly because of an increase in the proportion of fragmentation tracks in events.

At the time of writing, expectations for progress in b-physics have passed onto the b-factories (PEP-II and KEKB) and Run II at the Tevatron. The b-factories have been able to make exquisitely high precision measurements, most notably in the area of b-hadron CP-violation physics. Most of the obstacles to b-event reconstruction faced at LEP and the analysis techniques developed to overcome them, are irrelevant at b-factory experiments since they produce $b\bar{b}$ bound states in isolation with no confusion from the fragmentation system of particles. However, b-factories do not produce b-hadron states heavier than the lightest B^+ and B^0_d mesons and so it has fallen on the Tevatron experiments, CDF and D0, to further our knowledge of the heavier B^0_s, b-baryon and orbitally excited B-mesons.

The Tevatron environment is in many ways more challenging for b-physics than was the case at LEP. The good news is that the production cross-section is huge at the Tevatron $\sigma_{b\bar{b}} = 50\,\mu b$ compared to only ~ 6 nb at LEP. From here on in however, the news is nearly all bad:

- The $b\bar{b}$ cross section is just 10^{-3} of the total cross section, making the use of triggers for interesting events obligatory.
- b-physics triggers (typically high P_T leptons or tracks with large impact parameters) must have their efficiencies evaluated and unbiased b-physics data is then only available from the event hemisphere opposite to that in which the trigger registered.
- Despite the large Tevatron collision energy of 2.0 TeV the $b\bar{b}$ cross section peaks strongly towards very low momenta (see Fig. 8.2) resulting in a b-hadron boost of only $\gamma\beta \sim 2$–4 (c.f. around 6 at LEP I).
- A further consequence of the low momentum environment is a weak jet structure where properties such as directional correlations become washed out and quantities such as rapidity, which rely on a reference direction, will not work as well as was the case at LEP.
- Events are very rarely back-to-back and in a significant number of cases the b-quark partner to the b-quark that fired the trigger is not present in the opposite event hemisphere. This reduces any gains to tagging performance from including

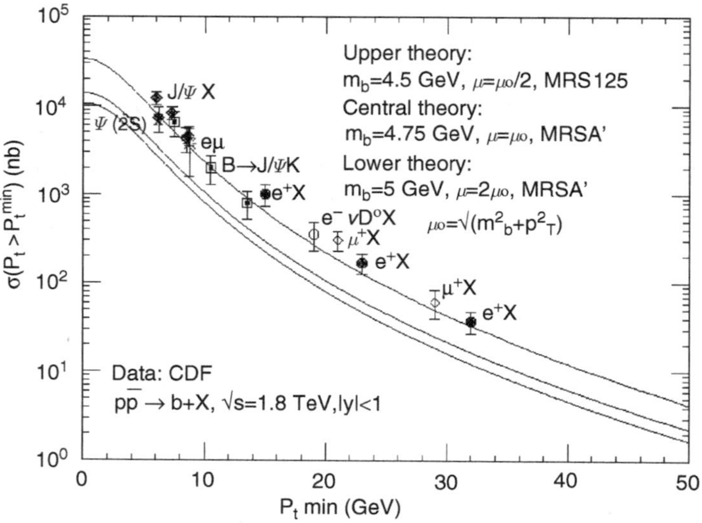

Fig. 8.2 Integrated cross section measurements for b-quark production from CDF compare to next-to-leading order QCD predictions. From [3]

information from the opposite hemisphere which was used with success at LEP (see Sect. 7.3).

- The ratio of tracks in an event originating from the b-hadron chain compared to background sources is found to be $\sim$ 3–4 : 1 compared to more like $\sim$ 1 : 1 in the LEP I data.

These aspects of the Tevatron environment degrade attempts to form quantities that are clearly correlated to the presence and properties of the underlying $b\bar{b}$ system. Despite this, LEP-style attempts to use more information in the Tevatron data than just high P_T-leptons for b-physics have met with quite some success. An example is the development of a LEP-style 'TrackNet' which uses a neural network to form a track-by-track probability for the track to have originated from a B-decay – analogous to the primary-secondary track probability P_{PS} described in Sect. 7.1.3. The variable relies heavily on a precise impact parameter measurement but also on variables derived from the transverse and longitudinal momentum characteristics of the track. Figure 8.3 shows the output and performance of the network probability on a sample of events triggered by a lepton with $p_T > 4\,\text{GeV}$ plus a track with a significant impact parameter (an SVT track). The track probability was used by CDF to boost the performance of the opposite side b-flavour tagging in the important first observation of B_s^0 oscillations in 2006 [5]. Following closely the development of weighted jet and vertex charge described in Chap. 5, CDF formed the following jet charge variable,

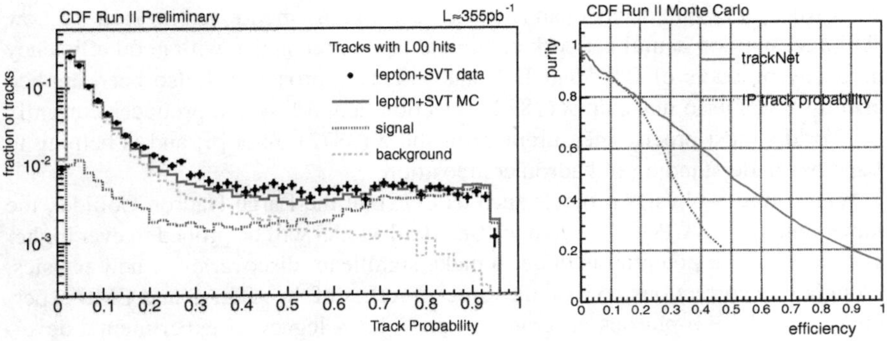

Fig. 8.3 (*left*) Output from the CDF B-track probability neural network where signal refers to tracks originating from the decay of a B-hadron ; (*right*) the performance of the track probability to tag B-decay tracks compared to the use of track impact parameters alone. From [4]

$$Q_{jet} = \frac{\sum Q_i \cdot p_{T,i} \cdot (1 + t_i)}{p_{T,i} \cdot (1 + t_i)}$$

where t_i is the track probability for track i. Weighting by the track probability in this way was found to improve the tag performance by about 25% and a distribution is shown in Fig. 8.4 for a sample of (μ + SVT track)-triggered data. A further example of LEP-inspired inclusive methods being applied in Tevatron data is in the

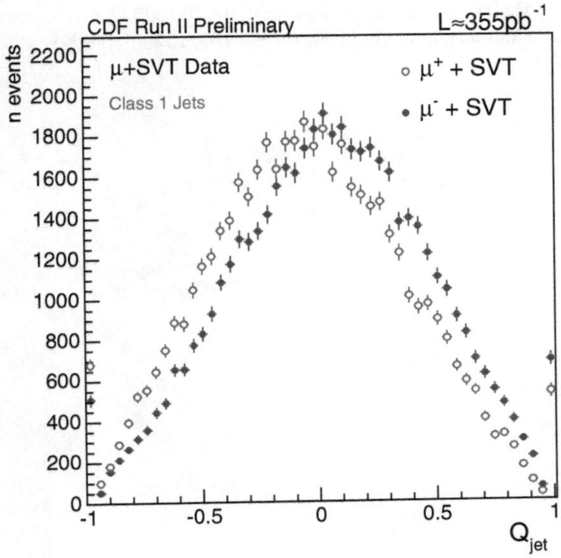

Fig. 8.4 Distribution of the track probability-weighted jet charge showing the shift visible between a $\mu^- + SVT$ and $\mu^+ + SVT$ triggered data sample. From [4]

reconstruction of rare decay channels. We have already mentioned in Sect. 7.2 how CDF have applied neural network combined tags to recognise with good efficiency the decay products of B^{0**} and B_s^{**} states. This approach has also been applied recently by CDF to make an $X(3872)$-selection network which produces currently the World's most precise measurement of the $X(3872)$ mass [6] and is helping to shape our understanding of hadron composition.

With future b-physics projects such as LHCb at the Large Hadron Collider, the implementation of CP-violation in the Standard Model will be probed to ever higher precision and the potential is there to make significant discoveries of new physics, through loop corrections, to rival the direct searches of the ATLAS and CMS experiments. The LEP b-physics program has provided a legacy of experimental developments and analysis techniques from which it is hoped the next generation of experiment will continue to benefit.

References

1. ALEPH Collab., S. Schael et al.: Eur. Phys. J. C **49**, 411 (2007); ALEPH Collab., R. Barate et al.: Eur. Phys. J. C **12**, 183 (2000); L3 Collab., M. Acciarri et al.: Phys. Lett. B **485**, 71 (2000); OPAL Collab., G. Abbiendi et al.: Phys. Lett. B **609**, 212 (2005); OPAL Collab., G. Abbiendi et al.: Eur. Phys. J. C **16**, 41 (2000); DELPHI Collab., J. Abdallah et al.: Eur. Phys. J. C **60**, 1–15 (2009)
2. ALEPH, DELPHI, L3 and OPAL Collabs.: Phys. Lett. B **565**, 61–75 (2003)
3. M. Paulini: Int. J. Mod. Phys. A **14**, 2791–2886 (1999)
4. G. J. Barker, C. Lecci: *Neural network based jet charge tagger*, CDF note 7482, available at http://www-cdf.fnal.gov/physics/new/bottom/bottom.html
5. CDF Collab., A. Abulencia et al.: Phys. Rev. Lett. **97**, 062003 (2006)
6. CDDF Collab., T. Aaltonen et al.: Phys. Rev. Lett. **103**, 152001 (2009)

Index

G.J. Barker, *b-Quark Physics with the LEP Collider*, STMP 236, 169–170,
DOI 10.1007/978-3-642-05279-8,